World Satellite Communications

and

Earth Station Design

World Satellite Communications

and
Earth Station Design

Brian Ackroyd, CEng, MIEE

CRC Press
Boca Raton   Ann Arbor   Boston

Contents

Preface

It is probably true that most people are aware of satellites and their use in communications, particularly TV, but there is very little awareness of the ground station aspect of a satellite system and the elements involved in the design of ground stations. In many books on satellite communications the emphasis is on one aspect of the system such as orbital utilisation or satellite station keeping, or a specific area of satellite communications such as VSATs. This book attempts to redress the balance between satellite and earth station information by concentrating on the parameters that affect the design of earth stations, even though the element may be the satellite design and control for example.

In addition to the emphasis on ground station elements, all the areas of application for satellite communications are considered, and the book is organised into separate chapters for specific areas of satellite use. It could be said that the current time is a watershed for satellite communications because of an explosion in digital communications and a coincident running down of analogue forms of transmission, which has led to the increased implementation of optical fibre schemes for long distance trunk communications. This growth of optical fibre links is said to be pushing satellite communications into niche areas of communications, but here too the book sets out to show that the basic premise is not correct and that the use of satellite communication systems has commercial and technological advantages over any other method, especially for certain uses such as remote sensing and mobile communications. The book therefore looks at all the major areas of satellite use and considers the parameters involved in practical system design.

The book provides details of the regulatory bodies that control the parameters of satellite systems and earth station design and this is particularly pertinent now when there is worldwide pressure towards deregulation of satellite services, partciularly in the field of television systems. These changes are matched by the changes in technology that are taking place in fields such as high definition television, which can only use satellite as the method of transmission.

The trends in all areas of satellite use are considered with the recognition that the major determining factor in choosing any communication system will not be the use of advanced technology but the cost effectiveness of the method. Satellite communications has always been an exciting area to be involved in and this excitement has not diminished even though the industry is now more than twenty-five years old. There are always new avenues to explore and problems to solve, and I hope that this book will convey some of the excitement and also the breadth of technology encompassed in a satellite system and the importance of the impact that satellites will make upon wide areas of our lives.

Finally I would like to thank three people who have assisted me greatly in producing this book: firstly the Commissioning Editor, Bernard Watson for giving me the opportunity to write this book and for all his assistance during the writing of it, secondly the sub-editor, Catherine Fear, whose advice and hard work made such a difference to the final version of the book and finally my wife, who was so patient during the long months that I spent at the word processor.

Brian Ackroyd

Chapter 1

A History of Satellite Communications

1.1 INTRODUCTION

The era of satellite communications began in October 1945, with the publication in *Wireless World* of an article by Arthur C. Clarke, which proposed a worldwide system of communications using three satellites orbiting the earth in a geostationary orbit. This vision took another 12 years to come to fruition when the USSR developed the technology that allowed the SPUTNIK satellite to be launched. It was a further five years before satellites began to be utilised as communications systems.

The first communication satellites were put into elliptical orbits that resulted in an orbit time of less than three hours; an example being Telstar which orbited the earth in 157 minutes. These early communications satellites required special earth station design to allow the satellite to be tracked while it was in the earth station's window of view. The first geostationary satellites, such as SYNCOM, were launched in 1963, to relay television from the 1964 Olympic Games.

The move towards setting up a worldwide satellite system came from within the USA which actually legislated in 1962 to create the Communication Satellite Corporation (COMSAT). This organisation was part private, part common carrier and part government controlled but had a policy objective of setting up a single global satellite communication system, even though at that time the satellite technology would only allow the use of elliptical orbits and required the use of very expensive earth stations. By 1965 the organisation had expanded to include 15 other countries, who formed the Interim Satellite Communication Organization and launched the first INTELSAT satellite named Early Bird; its communication capability was only 240 voice circuits and one television channel. Since its inception the INTELSAT system has expanded through six phases of satellite and now encompasses 112 countries. There are 16 satellites in geostationary orbit over the Atlantic, Pacific and Indian oceans, with the latest satellites having a capacity of 30 000 voice circuits and three television channels; in addition to the improved communication capability the satellite life has improved from one and a half years to ten years.

INTELSAT not only provides the satellites for the global system but also operates eight telemetry, tracking and command stations (TT and C) throughout the world. The information gathered by all of these stations is relayed back to the INTELSAT Operations and Spacecraft Control Center in Washington DC. This operations centre has the total responsibility for the maintenance of all the INTELSAT satellite positions and orbital inclinations, a list of satellites being shown in Table 1.1.

Table 1.1   INTELSAT satellites

Satellite position	INTELSAT type	Status
307° E	5/F3	IBS
310° E	4/F1	Spare
325.5° E	5/F4	Major path
329° E	4A/F1	Spare
332.5° E	5/F11	Spare
335.5° E	5/F10	Primary
338.5° E	4A/F4	Spare
341.5° E	5/F6	Major
359° E	5/F2	Spare
57° E	5/F1	Primary
60° E	5A/F12	Major
63° E	5/F5	Major
66° E	5/F7	Primary
80° E	4A/F6	Primary
174° E	4A/F3	Spare
179° E	5/F8	Spare

The definition of terms is:

Primary path   Operates within a region carrying minor routes;

Major path 1   Carries heavy traffic between major users (diversity);

Major path 2   As 1 but used as backup;

Spare   Used as diversity path;

IBS   For IBS only.

1.2   REGIONAL SYSTEMS

1.2.1   Introduction

The INTELSAT system was set up to provide a worldwide system, and can target a region if so required. There have been several regional systems set up to concentrate the service provided on either a country, a continent or a

Table 1.2 Major regional satellite systems

System	Region	Orbit position
EUTELSAT	Europe	10° E
AUSSAT	Australia	156° E, 160° E, 164° E
ARABSAT	Middle East/N Africa	19° E, 26° E
BRAZILSAT	Brazil	65° W, 70° W
Morelos	Mexico	113.5° W, 116.5° W
INSAT	India	74° E, 94° E
Palapa	Indonesia	108° E, 113° E, 118° E
Telecom	France	8° W, 5° W, 3° E
Anik	Canada	104.5° W–117.5° W

defined region. There are rules to define the setting up of satellite systems outside INTELSAT and these rules have been devised to prevent traffic being taken from the international business of INTELSAT. The major systems are shown in Table 1.2.

1.2.2 EUTELSAT

The European Telecommunications Satellite organisation (EUTELSAT) was first considered in the early 1970s as a means of linking the national networks of those countries that belonged to the European Space Agency. The actual planning of the system was begun in 1977 with the formation of Interim EUTELSAT, which was to manage the European Communication Satellite (ECS). It was intended that the system would cover an area bounded by Scandinavia in the North, Ireland in the West, North Africa in the South and Yugoslavia/Turkey in the East.

The organisation that operates and controls the system is structured in a similar way to INTELSAT, in that a number of countries participate in the EUTELSAT organisation but EUTELSAT plans, implements and operates the overall system. At the beginning it was decided that the EUTELSAT system would utilise a new frequency band, Ku Band in order to avoid interference with INTELSAT and to ensure that capacity of system could be maximised. This new frequency band was very much an unknown in terms of system performance and earth station design and it was decided that, prior to the main satellite being put into place, there would be an experimental phase which would determine both the satellite and the earth station requirements; to this end the experimental Orbital Test Satellite (OTS) was launched in 1978 into an orbit at 5° E, the whole experiment being supervised by the European Space Agency.

The ECS system comprises two phases of satellite design and the list of satellites is shown in Table 1.3.

Table 1.3   EUTELSAT satellites

Satellite	Orbit	Launch
EUTELSAT 1/F1	13° E	1983
EUTELSAT 1/F2	7° E	1984
EUTELSAT 1/F3	10° E	1985
EUTELSAT 1/F4	7° E	1987
EUTELSAT 1/F5	7° E	1988
EUTELSAT 2/F1	7° E	1989
EUTELSAT 2/F2	7° E	1990
EUTELSAT 2/F3	7° E	1991

The countries participating in EUTELSAT are shown in Table 1.4.

Table 1.4   EUTELSAT participating countries

Austria	Malta
Belgium	Monaco
Cyprus	Netherlands
Denmark	Norway
Finland	Portugal
France	San Marino
Great Britain	Spain
Greece	Sweden
Iceland	Switzerland
Ireland	Turkey
Italy	Vatican City
Liechtenstein	West Germany
Luxembourg	Yugoslavia

The initial system was designed to interlink all of the EUTELSAT participating countries, who were also members of the Conférence Européenne des Administration des Postes et des Télécommunications (CEPT – or European Conference of Postal and Telecommunications Administrations). Each member was to provide a single point of entry into its national network, which would carry telephony, data and telex traffic between CEPT countries and television between all members of the European Broadcasting Union (EBU). It was therefore necessary to make sure that the North African countries could receive the television transmissions.

EUTELSAT services

Television transmission
The major business, generated by the ECS system, has been in the distribution of television signals between member countries. In general the service

has been associated with cable networks and Satellite Master Television (SMATV) systems involving the full-time utilisation of a satellite transponder. It should be noted that the physical carriers of the service were the approved PTT organisations in the individual countries and it was not until 1989, when deregulation of the carrier provision began in Europe, that operators other than PTTs, were allowed to provide uplink facilities.

The second area of use to which the TV transponders are put, is occasional television transmissions such as sporting events and important public events.

In addition to the above uses EUTELSAT has a special relationship with the European Broadcasting Union (EBU) in that until 1994 it will provide the EBU with full transponder leases; these transponders allow the EBU to provide a high quality service across Europe and North Africa.

The characteristics of the various services are shown in Table 1.5.

Table 1.5   EUTELSAT EBU TV services

Main station		
Video signal/noise	53 dB	99% of month
Carrier/noise	15.5 dB	
Sound carrier signal/noise	56 dB	
Bandwidth	36 MHz	
TVRO		
Video signal/noise	50 dB	
Carrier/noise	13 dB	

The frequency of operation is in the 10.95–12.75 GHz band.

Business communications

An important element in EUTELSAT's future planning has always been the provision of a data service for the business community and it established, from the beginning, a Satellite Multiservice System (SMS) that allowed the user the opportunity to have a small earth terminal on their premises close to the terrestrial interface. The services offered are:

(1) Speech, data or compressed video data transmissions;
(2) Point to point communications;
(3) Point to multi-point communications.

The above services can be provided on a unidirectional or bidirectional basis; in addition the system has been designed to operate in both an open or closed network mode. In an open network all operators must conform to the same specification thus ensuring that there is full interconnectivity between all parts of the network. In a closed network mode of operation the users may deviate from the specification for the earth station in order to maximise the performance for their particular requirements. The SMS system utilises two forms of transmission; SCPC which can provide bit rates of up to 2048 kbits and TDMA which is designed to provide a high quality data transmission service and for EUTELSAT II it operates in a demand assigned mode. The TDMA characteristics are:

Modulation	2–4 Phase PSK
Bit rate	24.576 Mbits
Encoding	Differential
Frame length	20 ms
Word length	32 symbols

Public service telephony

In the public service sector of the EUTELSAT system the information is carried digitally and each earth station accesses the earth station in a predetermined pattern or time sequence.

1.2.3   AUSSAT

Introduction

A satellite communication system is ideal for Australia which has vast distances to cover and long distances between major centres of population. The system was put into operation in 1986 after more than ten years of planning by the Australian Government's National Satellite Task Force which considered a system that would cover Australia, New Zealand and Papua New Guinea; in the event New Zealand opted out of the initial system considerations, resulting in the system design, covering Australia and Papua New Guinea, being completed in 1980 and requests for quotations being issued.

The system uses three satellites situated at 156,160 and 164 degrees east in a geostationary orbit, with all three satellites being identical in performance. The satellites operate at Ku Band, uplinking between 14.0 and 14.5 GHz and downlinking between 12.25 and 12.5 GHz. Each satellite has 15 transponders each with a bandwidth of 45 MHz.

The satellites at 156 and 160 degrees east provide coverage of the whole of mainland Australia, including Tasmania, the third satellite, at 164 degrees east, provides coverage of the South West Pacific Basin. It should be noted that the third satellite also provides coverage of New Zealand, who now leases transponders from AUSSAT.

AUSSAT has had eight major city earth stations (MCES) constructed, as shown in the Table 1.6.

Table 1.6   AUSSAT city earth stations

MCES	Antennas (Number)	States (Size, m)	
Adelaide	2	13	
Brisbane	1	18	
Canberra	1	13	Unattended
Darwin	2	18	
Hobart	1	13	
Melbourne	1	13	
Perth	2	13	Attended
Sydney	2	13	

The whole network is controlled from the AUSSAT Network Operations Centre in Sydney. The only manned stations are in Perth and Sydney, the remainder being monitored from Sydney. The AUSSAT system is designed to provide a wide variety of options to suit the Australian circumstances; services are in unidirectional or bidirectional modes and are generalised below:

(a) Full or part-time video and audio transmissions: these transmissions can conform to BMAC or PAL standards. The PAL system follows the CCIR Rpt 624–1, with PAL B having pre-emphasis to CCIR Rec 405–1 and a baseband interface CCIR Rec 270–2. The PAL B signals are carried on an FM transmission with the sound using an FM sub-carrier. The BMAC video signals are transmitted using FM but the sound and other data are time division multiplexed at baseband. The BMAC characteristics are defined in document HACBSS Standard 511. This standard relates to the Direct Broadcasting Community Satellite Service known as Homestead and Community Broadcasting Satellite Service (HACBSS), which is a service operated by the government-owned Australian Broadcasting Corporation (ABC) and is intended to provide high quality TV programmes to both city and rural areas. The system uses four 30-watt transponders per satellite operating regional spot beams with services deriving from Sydney, Perth, Brisbane and Darwin.

The reason for choosing the BMAC type of transmission, which is a proprietary system devised by *Scientific Atlanta*, is that it allows a considerable reduction in received signal level for the same quality picture level.

(b) Long distance telephony: using SCPC/CFM transmissions to cover the whole of Australia with a global beam.

(c) Special Service Broadcasting (SBS): for educational purposes to isolated students, effectively replacing the existing shortwave radio system currently in use. It utilises two transponders to transmit radio and television.

(d) Q–Net: the Queensland telecommunication network, which uses a leased transponder to provide statewide television and data services for educational, medical, video/audio conferencing and business and computer communications.

(e) ITERRA Network: run by the Australian Telecom, this will provide dedicated business and data communications across Australia. It will rapidly implement a more reliable service nationwide as the Aboriginal word ITERRA, 'to be quick', implies.

1.2.4 ARABSAT

The Arab Satellite Communication Organization was formed in 1976 with the intention of improving the regional communication of the Arab world and complementing the terrestrial and INTELSAT services.

The satellites, designed by Ford Aerospace, were finally launched in 1985 in orbits at 19° E and 26° E. The satellites had the characteristics shown in Table 1.7.

Table 1.7   ARABSAT satellite characteristics

Type	Three-axes stabilised
Frequency bands	Rx 5925–6425 MHz ⎤
	Tx 3700–4200 MHz ⎦ C Band
	Rx 2500–2690 MHz
Transponders	25 in C Band
	1 in S Band
Transponder bandwidth	33 MHz
eirp	31 dBW nominal (34–26) C Band
	41 dBW at beam edge S Band
Capacity	8000 telephony/7 TV channels

The system can provide telephony, television, data, telex and telegraph services and uses various types of earth station:

(a) Earth stations operating between large metropolitan areas. These handle a number of telephony channels representing the trunk routes of the system. They can also originate television transmissions and receive them.

(b) Earth stations operate between the smaller towns and cities. Their needs are met by the use of SCPC transmissions. The stations are capable of receiving but not transmitting television.

(c) Earth stations for emergency communications which use transportable units that have smaller antennas mounted on some form of trailer. The purpose of the earth station means that it must be capable of rapid deployment, within a few hours of any situation arising. The system was originally intended for voice only but TV is also envisaged.

(d) Community television utilising S Band frequencies, which will provide remote communities with educational opportunities, entertainment and information.

The total capability of the system is shown in Table 1.8.

Table 1.8   ARABSAT transponders

Route	Number of transponders	Type of Tx
Trunk	10	FDM/FM
Thin	2	SCPC/CFM
Regional TV	1	FM
Community TV	1	FM
Emergency	3	FM

1.2.5   BRAZILSAT

The physical size of Brazil lends itself well to satellite communications and for some years it leased transponders from INTELSAT, to provide telephony, and television transmissions nationwide. Brazil now has its own satellite system, operated by SBTS, using satellites designed by SPAR Aerospace of Canada and deployed at 65° W and 70° W. The basic characteristics are shown in Table 1.9.

Table 1.9   BRAZILSAT characteristics

Frequency	Rx 5925–6425 MHz
	Tx 3700–4200 MHz
Transponders	24
Bandwidth	36 MHz
eirp	36 dBW (Brazil)
Capacity	12 000 telephony or 24 TV Channels

The services to be provided by BRAZILSAT are:

- Distribution of TV programmes;
- Relaying of radio programmes using analogue SCPC;
- Telephony;
- Direct broadcast of TV programmes.

1.2.6   Morelos

The Mexicans originally used leased transponders to provide them with
countrywide communications. The first transponder to be used was a C
Band system that provided them with four television channels that could
relay TV to remote communities. The country has always had a strong link
with the USA and a network of cable systems has been developed in Mexico
taking feeds from the USA network. This means that there is a reasonable
infrastructure for distributing satellite programming.

The Morelos system is unusual in that it uses dual band satellites that
operate at both C Band and Ku Band. This allows great system flexibility in
that the C Band can be used with the already existing ground station
infrastructure, derived from the leased transponder operation, while the Ku
Band can be used for data services as well as standard telephony. The
satellites, designed by Hughes, have the characteristics shown in Table 1.10.

Table 1.10   Morelos satellite characteristics

Frequency band	C Band	Rx 5925–6425 GHz
		Tx 3700–4200 GHz
	Ku Band	Rx 14 000–14 500 GHz
		Tx 11 700–12 200 GHz
Transponders	C Band	12 at 36 MHz/6 at 72 MHz
	Ku Band	4 at 108 MHz
eirp	C Band	36 dBW at beam centre
	Ku Band	48 dBW at beam centre

1.2.7   INSAT

The Indian National Satellite (INSAT) network is the product of considerable development by the Indian Government which has made India one of
the leading nations in the field of satellite communications. The network uses
an indigenous satellite design. The system became operational with the
successful launch in 1983 of INSAT 1B into an orbital position at 74° E
(INSAT 1A was launched in 1982 but never became operational due to
incorrect orbit control). The satellite characteristics are shown in Table 1.11.

The INSAT system was the first to combine three types of services;
telecommunications, television and meteorology. The telecommunications
system provides the long distance links between PTT switching centres, these
being based on FDM/FM and FM/SCPC modes of operation. The television broadcast services are centrally originated in each region and transmitted at C Band but received at S Band to provide national programmes as
well as educational services. Meteorological services use a frequency of
402.75 MHz for data uplinking and one at 4034.55 MHz for downlinking.

Table 1.11   INSAT satellite characteristics

Frequency band	C Band	Rx 5935–6425 MHz
		Tx 3710–4200 MHz
	Data collection	402.75 MHz
	S Band	2515–2595 MHz
Transponders	C Band	12
	S Band	2
Channel bandwidth	C/S Band	36 MHz
	Data	200 KHz
eirp	C Band	32 dBW min
	S Band	42 dBW min
	Data	19 dBW/multiple carrier
Capacity	12 000 voice/data channels or 2 TV	

In addition to the major facilities listed above the satellite is used for standard time and frequency transmissions as well as All India Radio centrally originated radio news programmes.

1.2.8   Palapa

The Indonesian Government initiated its own satellite system, to provide communications across the whole of the archipelago, by setting up Perusahaan Umum Telekomnika (PERUMTEL), an agency that was intended to operate the complete satellite system. The actual system was inaugurated in 1976 with the launch of Palapa A01, a satellite designed by Hughes, which not only covered the whole of Indonesia but also provided services to all of the ASEAN countries. The system now has the satellites shown in Table 1.12.

Table 1.12   Palapa satellites

Satellite	Orbit	Launch
A01	83° E	1976
A02	77° E	1977
B1	108° E	1983
B3	113° E	1984*
B3	118° E	1986

*Failed to reach orbit.

The satellites operate at C Band and provide the total capability shown in Table 1.13.

The whole system is controlled from the capital Jakarta.

Table 1.13   Palapa satellite characteristics

Transponders	A01	A02	B1	B3
Number of transponders	12	12	24	24
Bandwidth MHz	36	36	36	36
eirp dBW	33	33	34	36
TWTA power W	5	5	10	10
Capacity (channels)	6000	6000	24 000	24 000
or TV	12	12	24	24

1.2.9   Telecom satellite system

The French domestic system was inaugurated in 1984 with the successful launch of the Telecom F1 satellite. This launch was the culmination of ten years of independent effort by the French which had its beginnings in the experimental 'Symphonie' satellites launched in 1974/75. The satellite system was intended to provide communications between metropolitan France, its colonies and former colonies.

The first satellite, F1, employs three frequency bands, as shown in Table 1.14.

Table 1.14   Telecom F1 satellite characteristics

(a) C Band	TF1	Occasional transmission to the Caribbean, Canada and French Guyana
	TF2	Digital telephone traffic between France and the Caribbean
	TF3/4	Non-video and voice capacity
(b) X Band		Military
(c) Ku Band	4T	EUTELSAT SMS leased high speed digital communications using TDMA at 24.576 Mbts
	2T	TV spot beam

The satellite provides semi-global and spot beams from a satellite position of 8° W, with a TWTA output power of 8.5 watts at C Band, 20 watts at Ku Band and 20 watts at X Band. The equivalent isotropic radiated power (eirp) transmitted is as below:

C Band	France	28.5 dBW	Semi-global
	Caribbean	35.0 dBW	Spot beam
	Guyana	26.5 dBW	Semi-global
Ku Band	France	47.0 dBW	

The second French satellite, Telecom F2, was launched in 1985, at 5° W, without C Band transponders in order to avoid interference with the INTELSAT satellites situated at 4° W and 1° W. Once again the X Band transponders are used for military purposes, operating at an eirp of 27.2 dBW. The Telecom spot beam is centred on France with an eirp of 50 dBW at the centre of the beam and covering the UK, Scandinavia, Eastern Europe and North Africa where the eirp at the beam edge is 41 dBW. The satellite is intended to provide an international business data service.

1.2.10 Canadian Anik service

The geographical size of Canada has meant that it has been in the forefront of communication systems using satellites, both as a user and exporter of satellite communications hardware.

Canada is not only a founding member of INTELSAT and INMARSAT, with links into the systems via Teleglobe Canada, a federal corporation, but it also has its own domestic system operated on a commercial basis by Telesat Canada which is controlled by a mixture of Federal Government and commercial interests. The first satellite, Anik 1, was put into orbit in 1972 with Anik 2 and 3 following in 1973, an unusual launch in that both satellites were co-located in orbit in order to gain the same effect as a single satellite. The A Series are now retired but have been replaced by the satellites shown in Table 1.15.

Table 1.15   Anik satellites

Satellite	Frequency	Modulation	Trans	Position
Anik B	C/Ku	SCPC/FM/TV	12/6	109° W
Anik C	Ku	SCPC/TDMA	16	107.5° W
				112.5° W
				117.5° W
Anik D	C	FM/TV	24	104.5° W
				111.5° W

1.3   SPECIALISED SATELLITE SERVICES

The ability of satellites to provide a global communication system has meant that a variety of specialised systems have been devised to cover maritime, and aeronautical, land and navigational systems and the development of these and their characteristics is dealt with in chapter 6.

1.4 SOVIET SATELLITE SYSTEMS

1.4.1 Introduction

The development of the Soviet satellite communication system has followed, in general form, that of the West, with initial systems being based on C Band technology with later developments being in Ku Band. In organisational terms the USSR established an international satellite system named INTER-SPUTNIK, which was open to any country. Table 1.16 shows the member nations:

Table 1.16 INTERSPUTNIK members countries

Afghanistan	Mongolia
Algeria	Nicaragua
Bulgaria	North Korea
Cuba	Poland
Czechoslovakia	Rumania
GDR	Syria
Hungary	USSR
Iraq	Vietnam
Laos	Yemen, PDR
Libya	

The system uses a series of Russian satellites, STATSIONAR, located at 80° E for Indian Ocean Region coverage and at 14° W for the Atlantic Ocean Region. These satellites utilise FM modulation to transmit television and FM/SCPC or PCM/QPSK for telephony. The frequency band for reception is 3700–3900 MHz and for transmission, 6025–6255 MHz.

The earth station has the characteristics shown in Table 1.17.

Table 1.17 STATSIONAR earth station characteristics

Earth station	G/T	31 dB/K
Antenna gain	Rx	52.0 dB
	Tx	54.0 dB
eirp	Voice	51.8 dBW
	TV	84.6 dBW
LNA noise temperature		60 °K

The Soviet satellite capability covers both geosynchronous and non-geosynchronous systems, using C Band, Ku Band, L Band and UHF frequencies. The satellite types are listed below.

1.4.2 Gorizont satellites

The original Gorizont satellites used C Band transponders which provided high power transmissions having TWTA powers of 15 watts for global and hemi-beams and for a spot beam a TWTA power of 40 watts. These levels of output power allow signal levels on global beam transmissions of 26–29 dBW and 42 dBW for the European spot beam. The detail is shown in Table 1.18 for the Gorizont system.

Table 1.18 Gorizont satellite characteristics

Rx Band	5925–6225 GHz	
Tx Band	3650–3950 GHz	
Ch BW	34 MHz	
Coverage	Spot	42.0 dBW
	Zone	31.0 dBW beam edge
	N/Hemi	28.6 dBW beam edge
	Global	25.4 dBW
Channels	5 of 15W	
	1 of 40W	

This type of satellite is used for voice, video and data traffic between the INTERSPUTNIK member nations. In addition it is used to relay the USSR domestic television service.

The Gorizont system has characteristics that mean that it is possible to receive its transmissions in Western Europe. The early days of satellite television in Europe were very oriented towards the USSR transmissions and required an earth station having the characteristics in Table 1.19.

Table 1.19 Gorizont TVRO earth station design

Parameter	Value
Antenna diameter	2.2 m
Antenna efficiency	75%
Antenna gain	38.0 dB
LNA noise figure	1.2 dB
LNA gain	50.0 dB
System noise temperature	3.6 dB
System G/T	4.5 dB/°K

The main problem in the reception of such transmissions has been in the demodulation of the sound signals.

The Gorizont satellites began to carry an experimental 11 GHz service in the early 1980s, which was used to test the characteristics of the 11 GHz

Table 1.20 Loutch satellite frequency plan

CH number	Fup (MHz)	Fdown (MHz)
1	14 025	10 975
2	14 075	11 025
3	14 125	11 075
4	14 175	11 125
5	14 225	11 175
6	14 275	11 475
7	14 325	11 525
8	14 375	11 575
9	14 425	11 625
10	14 475	11 675

components for earth stations etc. The 11 GHz system is called Loutch and uses the frequencies shown in Table 1.20.

1.4.3 Raduga satellites

The first Raduga (Rainbow) satellite was launched in 1975 into an orbit at 85° E and since that time over 15 satellites of this type have gone into service, providing voice and data transmissions using SCPC/FM for the transmissions and TDMA as the access method. The satellites operate in C Band, using a frequency range of 3425 MHz to 3925 MHz for the receive band and 5750 MHz to 6250 MHz for the transmit band using RHCP for its operation. The satellites use both global and spot beam transmissions with nominal level of the global beam being 30 dBW and a 45 dBW level for the spot beam. It is similar to the Gorizont system but does not have a spot beam facility at 11 GHz and uses the lower portion of the C Band for transmissions.

1.4.4 Ekran satellites

These satellites were allocated to the USSR at WARC 71, for use as a direct broadcasting satellite system. The actual satellite is similar to the Gorizont design though the payload sub-systems are different. The Ekran satellite is situated at 99° E and serves an area of 3.6 million square miles, though its design life is quite short, only two years. The satellite transmits a signal in the frequency band 706–726 MHz at a signal power of approximately 53 dBW. The uplinking of the TV and radio broadcast is from Moscow, using a carrier frequency of 6.2 GHz and utilising an earth station antenna diameter of 3.5 m.

1.5 SATELLITE PACKET COMMUNICATIONS

1.5.1 Introduction

There has been a considerable amount of work, both theoretical and experimental, in the use of satellites as wide area packet network (WAN) distribution points. This network could be implemented using VSAT terminals and with suitable multiple access protocols could give improved performance over conventional VSAT networks. One particular area of development has been in the use of split reservation upon collision (SRUC) protocols, first proposed by Borgonovo and Fratta in 1978. SRUC working allows the system to operate with a slotted ALOHA protocol provided that no collision occurs; once a collision takes place and is detected by the system the system changes to a reservation protocol which uses a TDMA reservation channel, giving a stable system, even when a reasonably high throughput of messages is taking place.

The reservation system, which can give random access, can either be implicit or explicit; in the implicit case the allocation is made according to use, the system monitors when each station is using a slot and that station will retain that slot until it no longer needs it. In the explicit case the system reserves a time slot for a specific number of messages.

The use of a slotted ALOHA protocol can be inefficient, in terms of channel utilisation, if a stop and wait slotted ALOHA scheme is used. This inefficiency is caused by the delay characteristic of a satellite link which puts 0.25 seconds waiting time into the decision. This can be overcome by a modification to the decision-making process, where the packets are sent continuously and also have a continuous ARQ system such as a 'go back to N', where the effect of a collision is to make the system re-transmit the collided packet and all subsequent packets as determined by the reservation protocol.

With the SRUC protocol the channel time consists of a number of frames which are themselves divided into large slots. The large slots, within the frame, are made up of a single data slot and a number of sub-slots. The data slot is used to transmit a fixed length packet while the sub-slots carry signalling information used on a fixed assignment TDMA system – the number of mini-slots being proportional to the size of the large slot and the number of users.

The portion of the channel that is usually used for data is normally in the contention state where a user transmits both data packet and signalling information in a sub-slot. The sub-slot information tells the system which are packets still waiting to be sent successfully. If a collision occurs then the sub-slot information tells the system to switch into the reservation mode. This means that the system is no longer able to operate in a random access mode, but each user must now send reservation requests in a sub-slot; the sub-slots from all users forming a queue, in the order in which they are received.

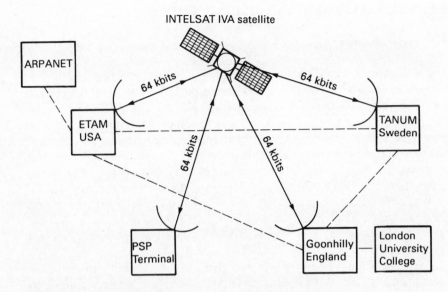

INTELSAT IVA satellite

Fig. 1.1 SATNET network configuration.

1.5.2   Atlantic Packet Satellite Network (SATNET)

General

One of the major experiments in satellite packet networks is the SATNET project which was an international. operation linking the United States, London University in the UK and the Norwegian Defence Research Establishment. The network was used as a vehicle for the testing out of a variety of network access and control methods, and can be represented as in Fig. 1.1. The main interest in this work is because it utilised and refined a new form of demand assignment which could operate both in packet or message mode. The particular access form is known as Priority-Oriented Demand Assignment (PODA) and this assignment method is used in two forms; fixed PODA (FPODA) for networks that have only a small number of users and controlled PODA (CPODA), which is used for a large system which may have variable traffic demand and patterns. The basic channel contains three types of information contained in two sub-frames:

(a)   An information sub-frame which is divided into two sub-frames; one associated with assignments from a centralised source and the other using a distributed control.

(b)   The centralised control information is relayed to all the participating users and the assignments can only be initiated once this has taken place. Due to the transmission delay of the satellite link the timing of the frame is scheduled one satellite link delay time ahead.

The distributed control information can thus give reduced delay compared to centralised control because the traffic assignments are made in the current frame being transmitted. The distributed system operates by every station in the network having the same assignment algorithm. Each user has the history of the channel usage over a single frame which allows the orderly use of the network by each station. The distributed system is likely to produce errors due to synchronisation difficulties at each of the stations.

SATNET transmission

The SATNET system uses a single satellite channel and all earth stations in the system transmit up to that channel on the 6 GHz carrier. All earth stations receive the satellite transmission at 4 GHz, i.e. the system works in a broadcast mode each station being given an address which precedes the data message. The frame structure is shown in Fig. 1.2. The frame is 128 virtual slots long, with a virtual slot being 10.24 ms long, giving a frame time of 1310.72 ms.

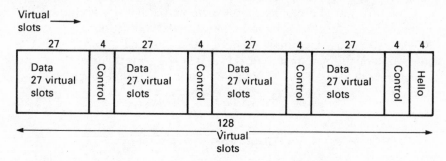

Fig. 1.2 SATNET overall frame structure.

The frame is divided into four blocks of 27 virtual slots each separated by a 4 virtual slot control slot. The final slot is used for hello messages which are transmitted to the four major stations.

The message format has the form shown in Fig. 1.3. The first 64 symbols are used for modem acquisition, the next 16 symbols provide for synchronisation and phase ambiguity resolution. The packet header has 128 symbols and the actual data message can be up to 1024 symbols, made up of 8-bit words.

One of the major elements of the SATNET earth station is the Satellite Interface Message Processor (SIMP) which provides the interconnection with the terrestrial network. The SIMP performs channel scheduling and synchronisation; forms the packets and controls the burst transmissions; in addition it also carries out the function of distributed control, thus obviating the need for a central control earth station. The transmission between major earth stations is QPSK.

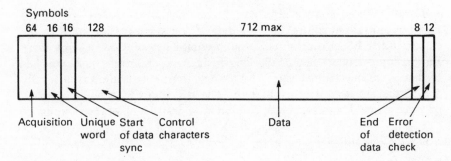

Fig. 1.3 SATNET message format.

The other unusual features of the system are the hello packets. These are transmitted by every station and received by every station. The packets are counted in the SIMP and are a measure of failure rate and also determine if the messages are to be re-transmitted.

The SATNET system is now operational rather than experimental, it is being used for test and monitoring of each packet and is a useful way of testing out access protocols in various transmission situations. The use of CPODA as a control algorithm has been found to be more useful for larger networks, though for four-station networks it does give some problems in contention detection. SATNET will be expanded with two more earth stations to check out packet networks operation and the access protocols.

1.6   USA DOMESTIC SATELLITES

1.6.1   Introduction

The technical performance of satellites and the use of the orbital arc is controlled by the FCC in the USA but there is no overall controlling entity in terms of access to satellite communications. The development of the satellite communication network has been on the basis of commercial requirements and, following deregulation, the growth of new common carriers. This environment has led to a very flexible forward looking industry which not only produced satellites but also a wide range of satellite earth stations. The initial systems utilised C Band and were the first to develop dual polarisation transmissions, effectively doubling the satellite capacity. Just as the early days of satellite communications, worldwide, were dominated by telephony requirements, the same applied to the USA. However from 1975 onwards the development of satellites as a television delivery system with immense results worldwide and later on with VSATs, led many companies to move from leased transponder operation to the launch of their own satellites; this applies to telephony, television and business data services.

The congestion of the orbital arc has taken satellite systems into the Ku Band of frequencies, though it should be noted that this band encompasses a different frequency range to that in Europe. The differences that obviously affect earth station sub-system design, such as antennas, LNAs and HPAs, thus lead to problems of inter-operability especially in the area of transportable systems. In chapter 9 it can be seen that there are different television standards which, although derived before the satellite era, produce the need for different earth station receiver design due to differences in sound sub-carrier frequencies etc. The modes of access are the same as those in Europe though the digital hierarchy differs and therefore special measures have to be taken when linking the two continents to provide interoperability between the USA standards at 56 kbits and the CEPT at 64 kbits.

1.6.2 USA satellite systems

The major USA satellite operators are listed in Table 1.21.

Table 1.21   Major US operators

Operator	Satellite series
Satellite Business Systems	SBS I–IV
Western Union	Westar
GTE	Spacenet/GStar
RCA	Satcom
Hughes	Galaxy
ATT	Telstar

Satellite Business Systems

Introduction
The SBS system was the first US attempt to provide a business satellite data delivery vehicle which utilised digital QPSK modulation controlled as a system by TDMA access. The venture was set up by IBM and Aetna Life and was a very ambitious project which attempted to break away from the more traditional mixed voice and data communications utilised by other systems. The satellite gives a continental US coverage (CONUS) beam, though SBS also acts as a distribution system for international traffic.

The original concept had to be modified because, although technically innovative, it did not generate sufficient business to be financially viable. This led to changes in ownership and business direction, with the addition of telephony services on a USA-wide basis. The latest satellite SBS IV is fitted with a number of spot beams that give sufficient power (50 dBW) to allow a satellite distribution system to be operated that is effectively a dbs operation. The basic characteristics of the SBS satellites are given in Table 1.22.

Table 1.22   Characteristics of SBS satellites

	1	2	3	4
Orbital position (°W)	99.1	97.0	95.0	91.1
Launch date	1980	1981	1982	1984
Satellite form		Spin Stabilised		
Frequency band Rx (GHz)		14.0–14.5		
Tx (GHz)		11.7–12.2		
Transponders	10	10	8/2	5/5
Transponder BW (MHz)		43		
Beam centre eirp (dBW)*	47	47	47/50	47/50
Third contour eirp (dBW)		41		
TWTA Power (W)	20	20	20/40	20/40
Services**	d	v/d/TV	v/TV	v/TV

*Eight channels at 47 dBW; two at 50 dBW
**voice/data/TV

The overall system can provide a full range of services required by modern businesses i.e. voice data, data (low and high bit rates), TV and video-conferencing. These facilities will be added to by SBS V and SBS VI, which will quadruple capacity with higher eirp to replace SBS I.

Westar satellite system

The Westar system is owned and operated by the Western Union Company who currently have a number of C Band satellites in orbit providing mainly TV and telephony services. The satellite series characteristics are detailed in Table 1.23.

Table 1.23   Characteristics of Westar satellites

	WIII	Westar WIV	WV
Orbital position (°W)	90.9	98.8	122.5
Satellite form		Spin Stabilised	
Launch date	1979	1982	1982
Frequency band Rx (GHz)		5.925–6.425	
Tx (GHz)		3.700–4.200	
Transponders	12	24	24
Transponder BW (MHz)		36	
Beam centre eirp (dBW)	32	39.75	34
Third contour eirp (dBW)	26	33.75	28
TWTA Power (W)	5	7.5	7.5
Services	v/d/TV	v/d/TV	v/d/TV

The Westar III is at the end of its design life which is ten years, though it is never an absolute cut-off point. The Westar system covers continental USA, Alaska, Hawaii, Puerto Rico and the Virgin Islands, acting as a carrier for a variety of customers using FM modulation and SCPC access.

GTE Spacenet

The GTE Spacenet system comprises three satellites providing a dual band fixed satellite service; the three satellites will eventually be amalgamated with three other Ku Band satellites to form a total complex known as GTE/-Spacenet. The system will act as a carrier for various companies that need CONUS coverage for services such as telephony and data transmissions. The satellite complex will be able to offer a flexible, easy to access service. The Spacenet satellite characteristics are shown in Table 1.24. The first two satellites are similar in terms of footprints and power levels, the third satellite departs from the others in that it provides east and west spot beams that provide Ku Band transmissions with a 10 dB higher power footprint compared to the CONUS pattern. The satellites provide vertically and horizontally polarised transmissions with two transponder bandwidths, narrow at 36 MHz and wide at 72 MHz using QPSK modulation and TDMA access systems.

Table 1.24 **Characteristics of Spacenet satellites**

	I	Spacenet II	III
Orbital position (°W)	120	69.0	88.5
Satellite form		3-axes Stabilised	
Launch date	1984	1984	1985
Frequency band Rx (GHz)		14.0–14.5	
Tx (GHz)		11.7–12.7	
Rx (GHz)		5.925– 6.425	
Tx (GHz)		3.700– 4.200	
Transponders		12 at 72 MHz/12 at 36 MHz	
Transponder bandwidth		72 MHz/36 MHz	
Beam centre eirp (dBW)			
Ku	43.8	44.6	49.0
C	37.9	38.1	–
Third contour eirp (dBW)			
Ku	41.8	42.6	47.0
C	36.9	36.1	—
TWTA Power (W)		16/8.5	
Services	v/d/TV	v/TV	

GStar satellite system

There are three satellites planned in the GStar series, two of which are operational with the third still to be launched. The satellites are unusual in that they not only provide CONUS coverage but embedded in that coverage are east and west spot beams. The details of the system are shown in Table 1.25.

Table 1.25   GStar satellites

	GStar satellites	
	I	II
Orbital position (°W)	103.1	105
Satellite form		3-axes Stabilised
Launch date	1985	1986
Frequency Rx (GHz)		14.0–14.50
Tx (GHz)		11.7–12.2
Transponders		16
Transponder bandwidth		54
Beam centre eirp (dBW)		
CONUS		46.4
W Spot		46.4
E Spot		47.9
Third contour eirp (dBW)		
CONUS		41.0
W Spot		41.0
E Spot		42.9
TWTA Power (W)	14 channels/20 W	2 channels/27 W
Services		v/d/TV

RCA Satcom system

The RCA system currently comprises five satellites, with a further two planned for launch by Ariane rocket or NASA shuttle; each satellite can provide different facilities allowing RCA American Communications to offer both civil and governmental agencies communication ability over an area much greater than CONUS coverage, reaching as it does Central America and the north of South America and as far west as Hawaii. The system relays radio and television to the American Forces and other organisations, such as Holiday Inn and RII. The details of the satellites are shown in Table 1.26. The five satellites operate at C Band with Satcom F5R, sometimes known as Alascom Aurora because it provides services between Alaska and the rest of the USA.

The method of providing the necessary services varies from satellite to satellite with FM being used for TV transmission, FDM/FM and SCPC/FM for telephony and SCPC/QPSK for data services. Satcom F1R and F4R are exclusively for TV programming, F2R accommodates MCI Sprint and the US Government requirements. The Aurora satellite differs from the other satellites in that it uses a solid state amplifier, rather than a TWTA, with consequent improvement in intermodulation distortion; this

Table 1.26   Satcom FR satellite characteristics

	II	III	IV	V	VI
Orbital position (°W)	72.0	131.0	82.0	142.9	139.0
Satellite form			3-axes Stabilised		
Launch date	1983	1981	1984	1982	1983
Frequency Rx (GHz)			5.925–6.425		
Tx (GHz)			3.700–4.200		
Transponder	24	6/18	6/18	24	24
Transponder BW (MHz)			36		
Beam centre eirp (dBW)					
CONUS	35	38	38	34.6	36
Alaska	32	–	–	34.6	38.3
Hawaii	–	28	–	36.1	–
Third contour eirp (dBW)					
CONUS	32	32	32	28.5	35
Alaska	29	–	30	28.6	–
Hawaii	–	22	–	6.1	–
TWTA Power (W)	8	8.5/5.5	8.5/5.5	8.5	8.5
Services	v/d/TV	TV	TV	v/d/TV	TV/d

improved performance allows the satellite operator to contain more channels in any transponder thus giving a more efficient and therefore less costly
performance.

Hughes Galaxy System

The Hughes Corporation is well known for its expertise in designing satellites and it is from this position that they became a common carrier, with
three satellites in orbit and a number in forward planning. The operational
satellite characteristics are listed in Table 1.27.

Table 1.27   Galaxy System satellites

	I	II	III
Orbital position (°W)	134	73.9	93.5
Satellite form		Spin Stabilised	
Launch date	1982	1983	1984
Frequency Rx (GHz)		5.925–6.425	
Tx (GHz)		3.700–4.200	
Transponders		24	
Transponder BW (MHz)		36	
Beam centre eirp (dBW)		36	
Third contour eirp (dBW)		34	
TWTA Power (W)		9	
Services	TV	v/d/TV	v/d/TV

ATT/Telstar

The logic of ATT having a satellite system to complement its terrestrial network is self-evident in that it not only gives increased capacity but can be used as a bypass network in the event of failure in any part of its normal network. The system uses three satellites to provide telephony and TV using FM modulation for TV and SCPC/FM for telephony. A data service known as Skynet 1.5 gives customers a 1.5 Mbit data service. The details of the ATT satellites are shown in Table 1.28.

Table 1.28 Telstar satellites

	301	302	3D
Orbital position (°W)	95.9	85.0	125.0
Satellite form		Spin Stabilised	
Launch date	1983	1984	1985
Frequency Rx (GHz)		5.925–6.425	
Tx (GHz)		3.700–4.200	
Transponders		24	
Transponder BW (MHz)		36	
Beam centre eirp (dBW)			
CONUS		·36.3	
Caribbean spot		36.3	
Hawaii spot		36.3	
Third contour eirp (dBW)			
CONUS		32.3	
Caribbean spot		32.3	
Hawaii spot		32.3	
TWTA Power (W)		8.5 W	
Services		v/d/TV	

The overall system not only provides CONUS coverage but has spot beams centred over the Caribbean and Hawaii which extend the service coverage beyond the United States.

1.7 SATELLITE COMMUNICATION REGULATION

1.7.1 Introduction

The history of satellite communications must also take account of the regulation of satellite communications over the period of almost 25 years. The regulation is firmly linked into the existing ITU regulatory framework, though other organisations now involved are more regional, such as CEPT in Europe and the FCC in the USA. The area of regulation, though changing, is essentially:

- Defining the frequency bands of a particular service;
- Defining the quality of the service when in operation;
- Setting the technical standards and parameters for the equipment to be used in the various services;
- Allocating frequency spectra to particular services in particular countries or regions;
- Co-ordination of terrestrial and satellite services to prevent excessive interference;
- Co-ordination of satellite systems;
- Allocation of geostationary arc;
- Tariffing principles;
- Interference from outside, non-communication sources.

The need for regulation is self evident for the following reasons:

- To prevent disruption of services;
- To permit easy interconnection between systems;
- To standardise on equipment interfaces and so prevent flexible operation of systems;
- To ensure that a non-renewable resource i.e. the radio frequency spectrum, is allocated fairly;
- To ensure that a scarce resource, i.e. the spectrum, is not polluted and is used as efficiently as possible.

1.7.2 Regulatory organisations

International Telecommunications Union (ITU)

This organisation is over 120 years old and is now the oldest telecommunications organisation in existence; it was founded in 1865 as a result of the growing use of telecommunications across national boundaries, though one can imagine that the work load was light considering that Guglielmo Marconi was not born until 1874. Today 164 member states belong to the ITU and until 1947 it was an independent organisation, at which time it became an agency of the United Nations, operating as an independent organisation under the umbrella of the UN. The structure of the organisation is shown in Fig. 1.4.

The major objectives of the ITU are set out in its Convention and can be modified at any Plenipotentiary Conference. These objectives are:

- To maintain and extend international co-operation for the improvement and rational use of telecommunications;
- To promote development of technical facilities and their most efficient operation and to improve the efficiency of telecommunications and increase their usefulness;

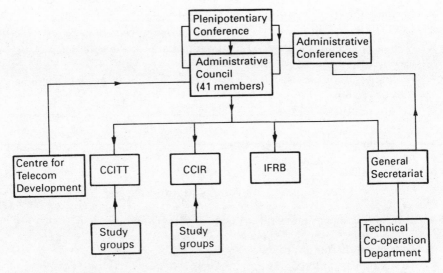

Fig. 1.4 Structure of the International Telecommunications Union (ITU).

- To harmonise the actions of nations in attaining these common goals.

The organisation operates in the following manner:

Plenipotentiary Conferences
Plenipotentiary Conferences comprise all of the 164 member states of the ITU; they are held every six years, where the Plenipotentiary must:

- Elect members to serve on the Administrative Council of 41 members;
- If necessary, modify the ITU convention;
- Determine the ITU budget, which is not a simple exercise but must take into account not only the finance for running the organisation but also the prevailing financial climate;
- Set general policy for the following six years.

Administrative Council
The Administrative Council meets each year to assess the work of the ITU and has the task of setting the budget each year within the framework of Plenipotentiary Conference decisions. In addition it reviews and sets the agenda for any of the ITU conferences that are being held in the coming years. The conferences are:

- World Administrative Radio Conference (WARC), which can theoretically cover a broad range of topics but usually due to pressure of work, confines itself to a single subject;
- World Administrative Telephone and Telegraph Conference (WATTC).

International consultative committees

There are two major consultative committees; CCIR which deals with radio regulations including satellites, and CCITT, which considers telegraph and telephone regulations. Each committee operates on a three/four year cycle of plenary assemblies, where the work of its study groups are considered; they have a Director, supported by a Secretariat and a number of study groups which varies depending upon the development of communication systems. The committee study groups are involved in producing three types of output:

- Reports on specific areas of communications;
- Answers to questions submitted by member countries;
- Recommendations on parameters for systems and equipments.

The CCIR study groups' results are contained in fourteen volumes, corresponding to the work of each study group (see Table 1.29).

Table 1.29   Work of the CCIR study groups

Volume	Subject
1	Spectrum utilisation and monitoring
2	Space research and radio monitoring
3	Fixed services below 30 MHz
4	Fixed satellite services/frequency sharing
5	Propagation in a non-ionised medium
6	Propagation in an ionised medium
7	Standard frequencies/time signals
8	Mobile services
9	Fixed services using radio relay systems
10	Broadcast services (sound)
11	Broadcast services (television)
12	Transmission of sound and television over long distances
13	Vocabulary
14	Indices

Although not directly involved with satellite communications the CCITT committee is directly related to equipment design because of the interface between the earth station and digital terrestrial services. It should also be noted that the structure and interaction of the two committees may well change in the future. The CCITT study groups are listed in Table 1.30.

The study groups meet at irregular intervals under the guidance of a chairman and work is carried out between meetings, often by PTTs such as British Telecom who have a considerable internal organisation dealing specifically with ITU activities. There are however new pressures on the ITU from three Regional Standards Committees viz CEPT, UST1 and the TTC Committee in Japan.

Table 1.30 CCITT study groups

Study group	Subject
1	Service definitions for telegraph, data etc.
2	Operation of telephone network and ISDN
3	Tariff principles
4	Transmission maintenance
5	Electromagnetic protection
6	Outside plant
7	Data communication networks
8	Terminal equipment for FAX etc.
9	Terminal equipment for telegraph
10	Languages and methods telecomms networks
11	ISDN and TN switching and signalling
12	Transmission performance of telephone NWs
15	Transmission systems
17	Data transmission over telephone networks
18	Digital networks including ISDN

Center for Telecommunications Development (CTD)
This centre became operational in 1986 and was created as an attempt to bridge the telecommunications gap between the developed and developing world; it was formed on the recommendation of a Commission set up by the ITU to consider the problem. The structure of the CTD is shown in Fig. 1.5.

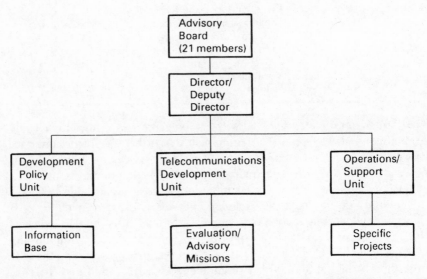

Fig. 1.5 Structure of the Center for Telecommunications Development (CTD).

The centre has been given a mandate to consider the following topics:

- Promote the importance of telecommunications and co-financing and collaboration on projects;
- Produce master plans for specific projects;
- Implement their specific development plans;
- Evaluate a country's communication needs;
- Advise countries on telecommunication solutions;
- Co-ordinate its activities with the Technical Co-operation Department

The CTD is funded on a voluntary basis by both government and private organisations; the DPU is the Center's database which supports the system design unit or TDU, and the OSU which is a pre- and post-support unit.

The International Frequency Registration Board
The IFRB reports directly to the General Secretariat and its function is to adjudicate on the new frequency allocations to decide if they conform to the ITU regulations. The IFRB has divided the world into three regions, which are:

Region 1   Europe, Middle East, Africa, USSR, Mongolia;

Region 2   N/S America, Greenland;

Region 3   India, Iran, SE Asia, Australasia, Japan, China, Pacific.

The frequency allocations are set for the various types of service and individual region. In Table 1.31 the service and frequency are defined.

In terms of regional frequencies the regions are allocated a number of the blocks shown in Table 1.31. This allocation is shown in Table 1.32.

The IFRB is also responsible for the daily management of the frequency spectrum, as applied to satellite communications and the various services to be provided worldwide. Every country submits applications for frequency allocations, which are checked by the IFRB to see if they conform to regional agreements as well as the ITU regulations; as part of this work the IFRB co-ordinates information and the current interference problems. The services detailed in Table 1.31 show that the various types share a frequency band, hence the need for the co-ordination of those services by the IFRB. Within the allocated frequency band there are protected and unprotected services as already explained; the IFRB is thus the co-ordinator of all shared services and has to take into account:

- Characteristics of the services;
- Maximum allowable interfering power level at the output of the interfered-with system;

Table 1.31

	Service		Frequency (MHz)
1	Fixed Mobile land Radio astronomy Space research	* * *	4990–5000
2	Radio location Space research Radio location	* *	5250–5255
3	Deep space research Amateur		5650–5725
4	Fixed satellite (EtoS) Radio location Amateur	* *	5725–5850
5	Fixed Fixed satellite (EtoS) Mobile	* * *	5850–5925
6	Fixed Fixed satellite (EtoS) Mobile Amateur Radio location	* * *	5850–5925
7	Fixed Fixed satellite Mobile Radio location	* *	5850–5925
8	Fixed Fixed satellite (EtoS) Mobile	* *	5925–7075

Note: The services that are marked with an asterisk are those that are protected, by the regulations, from interference from other services, while the others are unprotected.

Table 1.32

Region		
1	2	3
1, 2, 3, 4, 5, 8	1, 2, 3, 6, 8	1, 2, 3, 7, 8

- Transmission path characteristics of all systems;
- Parameters of the actual communication systems such as antenna gain, modulation etc.

The CCIR and CCITT Committees produce hypothetical reference cir-cuits or HRCs for both analogue and digital systems. The factors highlighted in these HRCs consist of allowable noise, video bandwidth, diversity operation, inter-satellite links, bit error rates and the effect of energy disper-sal. The ITU recommendations are detailed in volumes of reports and recommendations arising from plenary sessions and detail not only the technical performance of earth stations but also the methods by which the co-ordination takes place between the satellite systems. The IFRB first considers the proposed and existing network against the HRC; any problems arising are then considered in technical terms, a process that is followed by negotiation to modify the proposed network in order to remove any inter-ference problems.

The method of interference reduction involves:

- Frequency planning to separate satellite transponder frequencies;
- Design of earth station and satellite antennas to shape them so as to prevent overlap;
- Allocation of different transponder power levels to compensate for interference;
- Change of system design;
- Restriction of earth station use as in minimum elevation angles. The major element in the consideration of earth station performance is the antenna, which is designed in terms of radiation pattern, main beam and sidelobe characteristics, backward gain, forward gain and noise tem-peratures. The sidelobe pattern is crucial, in that the sidelobe gain directly affects the minimum satellite spacing. The antenna back radia-tion is important when considering the amount of interference with the terrestrial services. The CCIR recommendations provide a reference radiation pattern, a pattern which is being constantly reviewed in order to improve orbit utilisation.

Conférence Européenne des Administration des Postes et des Télécommunications (CEPT)

This is essentially a European organisation and was set up for the purpose of harmonising communications between European countries, including Eastern Europe, though to date only Yugoslavia has become a member. CEPT came into being in 1959 and now has 26 members and is in many ways organised in the same way as ITU in that it uses working groups co-ordinated by a harmonising committee. CEPT considers two areas, tele-communications and postal, each of which has a committee. The Tele-communications Committee has ten working groups. The working group of most interest to the satellite communicator is GT3 on radio communications and this group has eight separate technical committees relevant to satellite communications. The ten working groups are shown in Table 1.33.

Table 1.33   CEPT working groups

Working group	Area of interest
GT1	Telegraph operations and tariffs
GT2	Telephone operations and tariffs
GT3	Radio communications
GT4	Television and sound transmission
GT5	General tariff principles
GT7	Services and facilities
GT9	Long-term studies
GT10	Data communications (terminal)
GT11	Switching signals and protocols
GT12	Transmission and multiplexing

1.7.3   Future regulatory work

The work of network co-ordination continues with the new fixed systems being added to the network. The major problem over the next few years will be the co-ordination of mobile services, both land and aeronautical, which poses considerable problems to a regulatory body due to the variability of both transmit and receive earth stations' positions. The allocation of frequencies to mobile services is also difficult due to a spectrum shortage. The current allocation for aeronautical and land mobile services is in L Band, with land-based systems being given 806–890 MHz.

There will be changes in the ITU itself, in its organisational structure and function, with the possible merging of CCITT and CCIR functions to take account of the increasingly integrated nature of communications.

Chapter 2
Synchronous Satellite Systems

2.1 INTRODUCTION

The majority of satellite systems utilise a geostationary orbit. This type of orbit requires a minimum of three satellites if global coverage is to be achieved. A satellite must be designed to provide very stable satellite positioning. To ensure the maximum use of the geostationary arc, it must have on board antennas that are designed to provide minimum interference from sidelobe energy with adjacent satellites, earth terminal equipment and terrestrial microwave links.

The satellite must provide its own power source; it must also ensure that it remains in the correct attitude towards the earth and that the service is not interrupted by solar eclipses and sun transmit outages.

2.2 SATELLITE ORBIT

The altitude of the satellite, necessary to achieve a geostationary orbit, is derived from the fact that time taken for the satellite to move round the earth must be equal to the time that it takes to rotate once on its axis; this time is known as a sidereal day, which is three minutes and fifty-six seconds shorter than a solar day.

The satellite positions radius, S, is measured relative to the centre of the earth and is thus derived from equation 2.1 below:

$$S = (r + h) \tag{2.1}$$

where r = satellite height relative to the equator
         = 35 784 km
   h = earth's radius at the equator
         = 6378.165 km

The radius for a circular orbit takes into account the length of a sidereal day and the gravitational effects of the earth on the satellite. In this case the radius for a geosynchronous orbit can be derived from

$$S = \left[\sqrt[3]{\frac{\sqrt{u} \times T}{2\pi}} \right]$$ (2.2)

where μ = earth's gravitation parameter = 1.40765×10^{16} ft³/sec²

T = satellite period in seconds = 86 164 seconds

For a sidereal day this gives a value of S of 42 162 km and an earth's orbit height, relative to the equator of 35 784 km.

2.3 SATELLITE COVERAGE

To ensure that the minimum number of satellites are used they have to be situated over the three ocean regions; Atlantic, Pacific and Indian, requiring 120-degree separation as shown in Fig. 2.1.

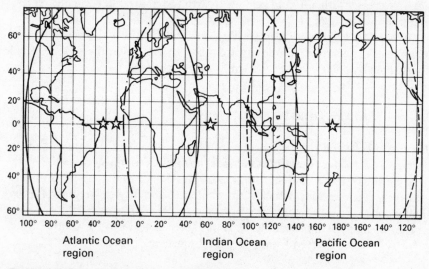

Fig. 2.1 INTELSAT global system coverage zone.

The satellite antenna beamwidth for full earth coverage, with a minimal system is 17.3°, which will illuminate 42.4% of the earth's surface, this area of illumination being called the satellite footprint. The form of footprint shown in Fig. 2.2 is known as a global footprint and although providing a maximum coverage it suffers from the disadvantage that the received signal at the earth's surface is relatively low due to the need to spread it so widely and it requires small aperture antennas on board the satellite which have a low isotropic gain.

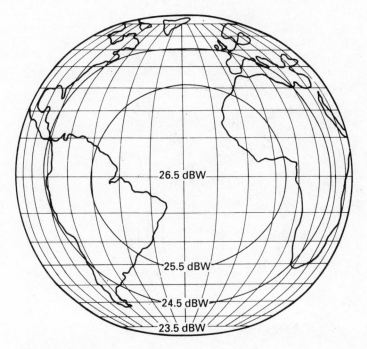

Fig. 2.2 Global beam coverage.

If the satellite system design is such that global coverage is not required then the available power can be concentrated into a much smaller area, thus providing a higher signal level at the earth's surface allowing smaller antennas to be used in the ground station. Other types of footprint are:

(a) Hemispheric, which covers a single hemisphere or 20% of the earth's surface.
(b) Zonal, which is restricted to a Continental zone.
(c) Spot beam, which concentrates the transmitted power over one country or area, as in a domestic satellite system. A typical footprint is shown in Fig. 2.3, which shows the signal level centred on a particular area, with the maximum signal level at the centre of the footprint and contour lines connecting points of equal signal as one moves away from the centre. The absolute level of signal can be shown in two ways; as a received power (dBW) or as an illuminating level (dBW/m²). The satellite footprints depict the expected signal levels at the earth's surface at the start of a satellite's life. This level will decrease with time as the satellite TWTA ages and reduces in output power.

The beam edge eirp also assumes that the TWTAs are operating at full output power and for the normal telephony and data services this is correct.

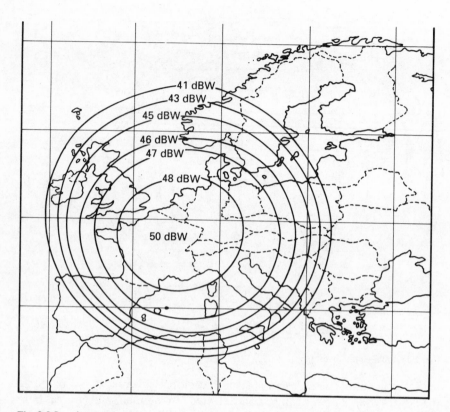

Fig. 2.3 Spot beam coverage.

However, for TV transmissions the TWTAs are operated at a power below maximum, i.e. backed off. The amount of back-off depends on the actual satellite and service and must be taken into account when calculating the received signal level.

2.4 ORBITAL GEOMETRY

It is true that for a given satellite position there is a signal loss, due to atmospheric conditions, as the signal travels to and from the earth station. This loss will vary depending upon the earth station position within the satellite footprint. In order to calculate the signal loss due to the attenuation of the atmosphere between the satellite and the earth station it is necessary to be able to calculate the actual distance, a subject that is dealt with extensively by Jacobs and Stacey (1970) and Jansky (1983). The actual computation is a matter of geometry as shown in Fig. 2.4.

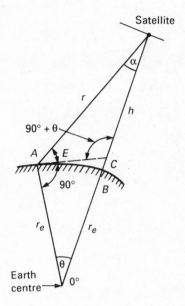

Fig. 2.4 Slant range, central, nadir, and elevation angles (geostationary orbit). r_e = earth's radius; h = satellite height; θ = central angle; α = nadir angle.

The graphical representation is shown in Fig. 2.5.

It has been shown that the maximum difference in distance from the satellite makes 1.32 dB difference in the received signal.

2.5 GEOSTATIONARY ORBIT UTILISATION

2.5.1 Introduction

The rapid growth of satellite communications has meant that considerable attention has been paid to orbit and spectrum utilisation. This need arises out of the fact that the geostationary arc is a finite resource which limits the number of satellites that can be in orbit at any one time. The physical limit means that particular attention has to be paid to satellite station-keeping, frequencies used by adjacent satellites and the satellites' transmission and reception characteristics.

The environment in which satellites operate is closely regulated by CCIR and by the satellite operators themselves, who not only have their satellites designed to meet the technical requirements associated with their allotted orbital position but also the design of each earth station is fixed and regulated by the satellite network operator.

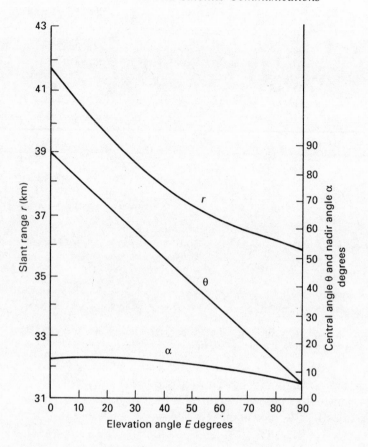

Fig. 2.5 Variation of slant range with elevation angle.

2.5.2 Satellite station-keeping

The ideal position for the satellite is one which does not vary at all, relative to other satellites and the earth station. This ideal would reduce satellite weight and size because it would not require thruster motors to reposition the satellite nor the fuel to drive the motors. In addition the earth stations would be less complex, requiring no tracking systems to follow the satellite movement and thus allowing simpler antennas at lower cost.

This ideal is of course not possible due to the variation of gravitational forces associated with the sun and moon which produce both long and short-term movements. The longitude movement is caused by the fact that the equator is actually slightly elliptical, meaning that the earth's gravitational field does not act directly towards the centre of the earth.

These short-term movements manifest themselves as a figure of eight relative to the earth station; the magnitude of this figure of eight being dependent upon the inclination of the satellite orbit and increasing as the square of the inclination angle. The peak latitude and longitude excursions relative to the sub-satellite point are given by:

$$\theta_{LAT} = \sin^{-1}(\sin I \times \sin x) \tag{2.3}$$

$$\theta_{LONG} = \sin^{-1}\left[\frac{\cos I \times \sin x}{\sqrt{1 - \sin^2 I \cos^2 x}} - x\right] \tag{2.4}$$

where I = orbital inclination angle

$$x = \frac{2\pi t}{24} \quad (t = \text{time in hours}).$$

The longitudinal motion relative to the orbital inclination is shown in Fig. 2.6.

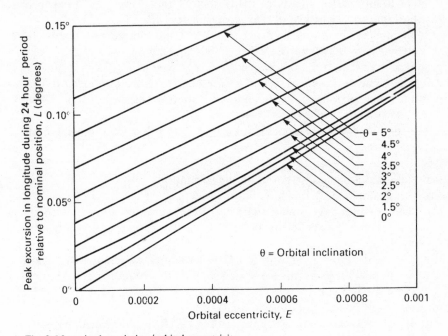

Fig. 2.6 Longitude variation/orbital eccentricity.

The gravitational effects actually change the angle of inclination of the orbital plane; in the equatorial plane this change of inclination varies from year to year between 0.75° and 0.95°. Fig. 2.6 also shows the effect of orbital eccentricity on the satellite position in longitude and it is obvious that very small eccentricities can cause considerable excursions about a mean position; these variations occur with time, even though initial orbital positioning can minimise the change as a cyclical effect due to solar pressure.

Gravitational effects can be minimised, given an accurate orbit, by placing the satellite in either of two positions; at 76.8° W longitude and 108.1° W longitude. However the earth's gravitational field decreases from these points in both directions thus producing satellite drift which is effectively a change in the orbit's period. This drift obviously determines the number of satellites that can be placed along the geostationary arc and thus has a prime effect on the orbital utilisation factor. Modern satellites can hold their position to 0.05° by means of improved initial positioning and the use of east/west station keeping motors.

2.5.3 Efficiency of orbit utilisation

Apart from the need to keep the satellite in an absolutely fixed position, relative to the earth, there are many factors that affect the number of satellites that can be placed on the geostationary arc.

In defining system usage there are two forms to be considered: first where all satellites use the same frequency bands and illuminate the same geographical area, i.e. they are homogeneous, and second a non-homogeneous case where one of these parameters is different. It is obviously easier to deal with the homogeneous case, though it is not the most efficient way to design a satellite system.

For the homogeneous model it is assumed that it has the following basic characteristics:

(a) Satellites are equally spaced on the geostationary arc
(b) An analogue or digital modulation system is used
(c) The satellites have similar characteristics in terms of equivalent isotropic radiated power (eirp), polarisations, frequencies and modulation methods.
(d) The earth stations conform to CCIR standards, especially with regard to the sidelobe performance, which should meet $29 - 25 \log \theta$ where θ is the off-axis angle.
(c) Propagation conditions are clear sky.
(f) Interference limits are as defined in the CCIR recommended limits.

The major elements that affect the utilisation of the geostationary orbit are:

Interference noise

The noise level that has been set for FM/FDM systems is 1000 pWOp, which is one tenth of the maximum value set for any fixed system. The spacing of the satellites varies inversely as the interference noise allowance changes, this change being very rapid about the reference level, although as the noise level approaches the maximum allowable level the law of diminishing returns applies.

Earth station antenna size

The achievable satellite spacing is a function of the earth station antenna size and is normalised relative to a unity value of $\dfrac{D}{\lambda}$ where:

D = Antenna diameter
λ depends upon the operating frequency.

2.6   SATELLITE CONSTRUCTION

The satellite is, to all intents and purposes, a microwave repeater which receives a signal transmitted from the ground, translates it to another frequency and amplifies it before transmitting the message to another destination. This function determines the form the satellite takes and the design of each portion of the satellite. The satellite must perform the following tasks:

2.6.1   Stabilisation

It must maintain its position relative to the earth in order to ensure that the same portion of the earth's surface is illuminated. This means that the satellite has to be stabilised in some way so as to prevent it from tumbling in its orbit.

There are two main forms of stabilisation:

(a)   The satellite is constructed in a cylindrical form, as shown in Fig. 2.7. This cylinder is spun about its axis, with a platform inside, that contains the active components, remaining static. The spinning action makes the satellite act like a gyroscope which retains its position relative to the earth.

(b)   The satellite can also utilise an active system of stabilisation, as shown in Fig. 2.8, operating in three axes to counteract roll, pitch and yaw. The satellite also uses rocket motors to maintain its position in orbit and for attitude control. These motors are attached to various points on the satellite and in modern systems use liquid propellants such as nitrogen tetroxide and mono-methyl hydrazine. The firing of these

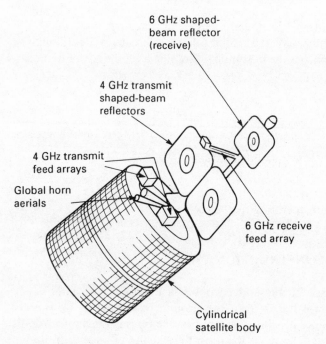

6 GHz shaped-
beam reflector
(receive)

4 GHz transmit
shaped-beam
reflectors

4 GHz transmit
feed arrays

Global horn
aerials

6 GHz receive
feed array

Cylindrical
satellite body

Fig. 2.7 INTELSAT IV satellite.

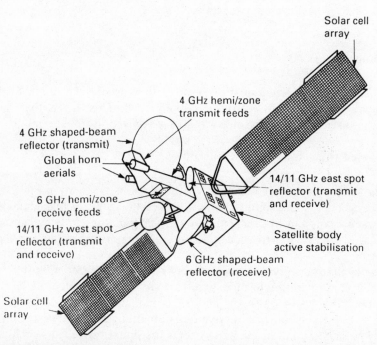

Solar cell
array

4 GHz hemi/zone
transmit feeds

4 GHz shaped-beam
reflector (transmit)

Global horn
aerials

6 GHz hemi/zone
receive feeds

14/11 GHz west spot
reflector (transmit
and receive)

14/11 GHz east spot
reflector (transmit
and receive)

Satellite body
active stabilisation

6 GHz shaped-beam
reflector (receive)

Solar cell
array

Fig. 2.8 INTELSAT V satellite.

motors is controlled from the ground control system which also controls the satellite during the launch phase and while operational.

2.6.2 Power

The satellite must be able to supply itself with power to operate the repeater systems and ensure that it remains fully operational during the whole of the year. For a spin-stabilised satellite the external cylinder is covered in solar cells, giving a solar array that has the equivalent area of a rectangle with sides $\pi. d$ and h where:

d = satellite diameter

h = satellite height

The total solar cell area is not illuminated all at the same time and is inefficient relative to the system used in the axes stabilised satellite, where two solar panels are deployed out from the satellite once the satellite is in position. The two solar panels are oriented towards the sun and are π times more efficient than the cylindrical system. The power system must operate over a period of at least ten years, taking into account the fact that solar cells are lost during the satellite's deployment due to meteorites etc. The power required is considerable, e.g. 3 kW for the ECS system. There is another effect to be taken into account in the design of solar cells, i.e. the effect of radiation. Solar radiation, especially during solar flare periods, reduces the power output and efficiency of each cell. The majority of the damage to the solar cells is caused by the protons emitted during solar activity which are at an energy level of 1–3 MeV. These high energy protons peneterate well into the individual solar cell and produce actual defects in the structure of the cell itself; this lowers the cell short circuit current and thus reduces cell output power.

The development of solar cells is thus not simply concerned with raising conversion efficiency, but is also with improving the cells' tolerance to radiation, in order to increase the useful length of life of the satellite's power system. Improvements have been made not only by changing the design of cells to a phosphorous on silicon construction, but also by providing the cells with a glass covering which reduces the radiation level arriving at the cell in proportion to the thickness of the glass. In addition the glass is coated with silicon monoxide, which is a reflecting agent, thus once again reducing the radiation level.

It has to be recognised that there are certain natural effects that cannot be avoided and must be taken into account in any satellite design e.g. solar eclipses. The effect of solar eclipses varies with the year as shown in Fig. 2.9.

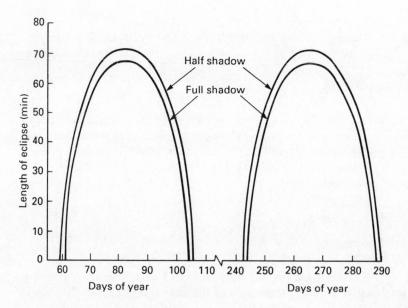

Fig. 2.9 Length of eclipse/day of year.

The maximum effects occur at the spring and autumn equinoxes, where the batteries must operate for 70 minutes a day at its worst, decreasing to zero twenty days on each side of the equinox. The backup batteries typically use nickel–hydrogen cells that are more reliable than the earlier nickel–cadmium type and are less affected by the thermal environment with a better charge and discharge characteristic. This type of system has a long life, is light weight and highly reliable. In addition to the batteries there is also a need for a power control system to ensure a stable supply voltage.

On the ESA 2 satellite there are two solar arrays consisting of four panels on each side of the satellite providing a total of 28 square metres of solar cells. The solar arrays are made up of a highly rigid carbon fibre honeycomb sandwich, a modern material giving considerable savings in satellite weight.

2.6.3 Communication system

The satellite must provide a communication system that can handle voice, data and television links; in essence the system is a number of earth stations operating in transmit and receive modes. The satellite sub-systems consist of parabolic antennas, often with multiple feed horn assemblies and beam forming networks; repeaters that receive the signal transmitted from the ground station, amplify it, downconvert and then demultiplex the signal. The demultiplexed signals are fed, via channel amplifiers, to travelling wave tube amplifiers, whose outputs are multiplexed after selection and then passed to the beam forming network of the transmit antenna.

2.6.4 Control system

The satellite environment must be controlled to ensure that all its subsystems are maintained within their design limits over the full life of the satellite. The control system is bound to be complex having, as it does, to take account of the temperature rise due to solar radiation and at the same time remove the power dissipated by the travelling tube amplifiers. The control system is both active and passive, though wherever possible passive devices are used, such as multilayer insulation and solar reflectors, with heat being conducted away from hot spots by the use of heat conducting pipes.

2.7 SATELLITE FREQUENCY

The selection of the transmission frequencies for satellite communications has to take account of the effects of the transmission media upon the signal. There are a variety of propagation medium characteristics that affect the signal, producing both first and second order effects. The system designer must take into account all these factors when considering both satellite and earth station design to ensure that best use is made of the technology available from a cost and performance point of view. The satellite payload weight has to be minimised and therefore the smallest and lightest subassemblies, such as high power amplifiers and antennas, must be used. In addition the balance between earth station and satellite size must be taken into account.

The frequency of operation must also consider the terrestrial frequencies that are in current use in order to avoid interference between the two systems.

2.7.1 Atmosphere attenuation

It is essential that the signal attenuation be defined as precisely as possible in order to minimise power amplifier, antenna and communication sub-system design; in this it differs from the design of terrestrial systems, where considerable design margins are allowed for fading and path attenuation uncertainties. The attenuation characteristics of the atmosphere are frequency dependent as shown in Fig. 2.10.

It can be seen that at certain frequencies the absorption of the signal increases rapidly. The major cause of attenuation is the absorption of

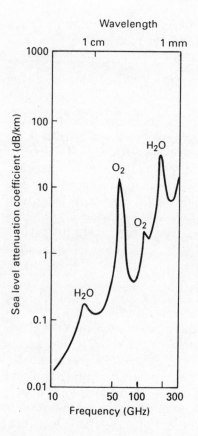

Fig. 2.10 Atmospheric attenuation/frequency in GHz.

microwave energy by oxygen, water vapour or rain. The major absorption frequencies are 22 GHz, 60 GHz and 119 GHz and the actual level of attenuation is a function of temperature, pressure and the actual composition of the absorbing medium. There has been considerable work carried out, on both a theoretical and practical basis, to derive attenuation characteristics especially with regard to molecular absorption which is the major cause of signal attenuation. The individual effects are:

Absorption due to gas and water vapour

The major atmospheric attenuation factors are oxygen, water vapour and precipitation. These have absorption lines in the millimetre wavelength region and losses are due to magnetic interaction with the incident field and produce absorption lines at 60 GHz and 118.8 GHz and are dependent upon temperature and atmospheric pressure. In the case of water vapour the vapour behaves like an electric dipole which produces attenuation due to interaction with the incident signal at 22.2 GHz, 183.3 GHz etc. The attenuation due to water vapour is dependent upon the path temperature profile, which is a slowly varying parameter but the other element, humidity level, varies rapidly due to both geographical location and the time of year.

The actual attenuation of a signal transmitted between satellite and earth station and vice versa is a function of path length, gas absorption constant and gas molecule distribution and is given by:

$$A_G = \int_0^r \gamma(h)\, dr \qquad (2.5)$$

A_G = attenuation in dB

h = height above sea level

γ = distance from earth antenna.

The National Bureau of Standards has produced an empirical relationship as shown below:

$$A_G = [1.4 + 0.09 f - 1.6\, e^{(-2.1f)}] \times [1 - e^{-(0.0054r)}] \times e^{-10\theta} \qquad (2.6)$$

where f = frequency in GHz

r = path length in km

θ = elevation angle

The above equation is said to apply to frequencies below 10 GHz.

Attenuation due to clouds and rain

There is a common element in clouds and rain which is water droplets. These water droplets, through temperature effects, can become snow or ice, each element being known as an hydrometeor. As the frequency increases so the attenuation increases; if at the same time the temperature decreases this will lead to an attenuation increase.

There has been considerable work done in investigating the effect of water droplets on microwave signals, both theoretical and practical. The attenuation due to rain is the major element in path delay and is derived from:

$$A = K \times p \times r \qquad (2.7)$$

where A = attenuation in dB/km

k = constant related to temperature, frequency, rainfall etc.

p = rainfall rate in mm/hour

r = distance between satellite and earth station

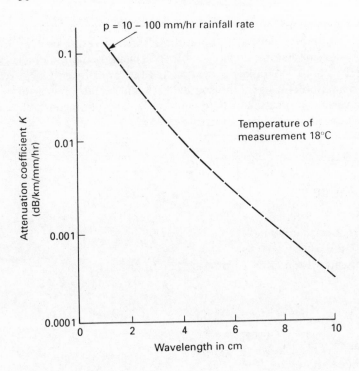

Fig. 2.11 Attenuation coefficient of rain drops/wavelength.

In order to derive the attenuation it is necessary to derive K, which is shown in Fig. 2.11.

The preceding discussion has concentrated on attenuation due to hydro-meteors, as these are major elements in atmospheric attenuation. There are, however, other minor effects due to dust or volcanic eruptions which can cause attenuation in the same way as rain, snow or ice. The extent of such attenuation is typically 0.2 dB/km and is dependent upon the moisture content of the dust.

2.7.2 Further propagation factors

In addition to losses caused by the propagation path the system designer must take into account a number of other factors which produce signal loss and degradation. These are:

Scintillation effects

The signal transmitted between earth station and satellite has to pass through a number of different layers of both troposphere and ionosphere; the action of these layers is to refract the signal and changes in the refractive index of the transmission medium produce an effect known as scintillation, apparent at the earth station as amplitude and phase variations of the signal. The effect of scintillation is especially apparent at low earth station antenna elevation angles and appears to the earth station as an apparent error in the antenna pointing angle, producing thereby an effective loss in received signal and is effectively an increase in the attenuation of the signal path.

In practical terms normal earth stations operate above an elevation angle of 5 degrees and thus scintillation effects represent a small effect compared to rain attenuation. The variations in amplitude of the received signal, due to scintillation effects, have a fading characteristic that varies depending upon the frequency of transmission, the geographical region, time of day and sun spot activity. The observed received signal variations in the Middle East, for instance, show peak to peak magnitude of over 10 dB, a change in input signal that produces severe difficulties in earth station tracking systems.

Faraday rotation

When the transmitted signal moves through the ionosphere it interacts with the earth's magnetic field to produce a rotation of the plane of polarisation of a linearly polarised signal; this effect is known as Faraday Rotation, an effect that is proportional to the electron content of the ionosphere. The amount of polarisation rotation is thus dependent upon the variations taking place in the ionosphere and therefore it changes from season to season, with sunspot activity, frequency of transmission etc. It is significant in C Band systems where compensation has to be made at the earth station but reduces in magnitude as frequency increases until at frequencies about 10 GHz it is not an important effect. The Faraday rotation effect has been found to be inversely proportional to the square of the propagation frequency and can in practice be partially compensated for by physical adjustment of the polariser on the earth station antenna, though it will not compensate for very large variations of the ionosphere due to sun spot activity or magnetic storms, where large scintillation effects can be observed.

Depolarisation and cross polarisation

In order to give system flexibility the transmission can be made on horizontally or vertically polarised signals, providing double the capacity at the same frequency. The use of polarisation diversity, as a means of increasing the system capacity, means that the depolarising and cross-polarising effects of the transmission medium have to be taken into account.

Depolarising occurs when the medium through which the signal passes causes the transmitted polarised wave to lose power in that polarisation, with the lost power not necessarily being orthogonal to the transmitted wave. Cross polarisation is, to an extent, a special case of depolarisation in that the lost power appears orthogonal to the plane of the transmitted polarisation. It is very important to note that cross polarisation results not only from atmospheric effects but also from the action of the receive and transmit antennas upon the signal.

There are two ratios that are of great importance in defining any system. The first, cross polarisation isolation, is the ratio of the co-polarised signal in a receiver channel to the cross-polarised signal in the same channel, assuming that each polarisation is transmitted at the same signal strength. The second ratio is cross polarisation discrimination which is the ratio of co-polarised received signal and the cross-polarised signal, when one polarisation only is transmitted.

The atmospheric causes of cross polarisation, for satellite communications, are rain and ice (cloud-borne). The level of cross polarisation is not only affected by the amount of rain but also the angle of the rain drops' major axis to the horizontal, known as the canting angle. The effect of the canting angle means that the use of horizontal and vertical polarisations is much more effective in combating the effect of cross polarisation, compared to circular polarisation. The effect itself is frequency and attenuation dependent and must take account of the elevation angle.

The cross-polar effect for hydrometeors is derived from:

$$X_{pd} = X_{pdr} - C_{ice} \tag{2.8}$$

$$= X_{pdr} - \left[X_{pdr}(0.3 + \frac{0.1 \log P)}{2} \right]$$

$$= X_{pdr} - \left[1 - (0.3 + \frac{0.1 \log P)}{2} \right] \tag{2.9}$$

where X_{pdr} = cross polarisation caused by rain

C_{ice} = ice crystal effect

P = proportion of the time for which X_{pdr} is not exceeded

In order to calculate the cross polar effect due to rain the following formula is used:

$$X_{pdr} = C_f + C_0 + C_i + C_c + C_a \tag{2.10}$$

where C_f = frequency dependent effect = $30 \log f$ for $f > 8\,\text{GHz} < 35\,\text{GHz}$
 $C_a = V \log(Ap)$ where $V = 20$ for f between 8–15 GHz
 $V = 23$ for f between 15–35 GHz
 Ap = path attenuation

$C_0 = -40 \log(\cos \theta)$ for $\theta < 60°$

$C_i = -10 \log[1 - 0.484(1 + \cos 4I)]$

$C_c = 0.052 \, \theta^*$ where θ^* is the standard deviation of rain drop canting angle distribution expressed in degrees

The above calculation can be scaled to another frequency and tilt angle by using the following formula:

$$X_{pdf_2} = X_{pdf_1} - 20 \log_{10} \left[\frac{f_2 \sqrt{1 - 0.484(1 + \cos 4I_2)}}{f_1 \sqrt{1 - 0.484(1 + \cos 4I_1)}} \right] \qquad (2.11)$$

This formula applies between 4 GHz and 30 GHz.

The above relationships have been compared to practical measurements which show that for superior performance it is better to use linear polarisation rather than circular. One practical result of these measurements has been to show wet snow on the earth station parabolic antenna causes far more cross polarisation than that due to propagation effects.

Atmospheric refraction

The passage of a signal through the various atmospheric layers cause a bending of the radio wave due to changes in electron density on the system path; this effect is seen when the signal is at a slant through the atmosphere and it is true to say that this effect is only noticeable at low elevation angles. Another refractive effect seen at low elevation angles, which causes path loss, is wave front incoherence which appears to reduce the antenna gain. This effect is frequency dependent and seems to affect large antennas with narrow beamwidths rather than small systems.

System noise

The 'goodness' factor of any satellite/earth station is defined by its gain noise temperature ratio. Those elements that make up system noise will be discussed in chapter 4; however one element of that noise is atmospheric noise, which is a non-controllable element in the equation, though the system design can mitigate certain effects.

The system noise injected by outside sources comes from four main effects:

(a) Cosmic background noise: this is a constant, injecting 2.8 °K into the system and comes from the Sun and a number of Star systems, such as Cassiopeia and the Crab Nebula. The level of noise varies with frequency, this variation being shown in Fig. 2.12. The level of galactic noise increases inversely with frequency below 1.6 GHz and is thus insignificant for most geostationary satellite systems.

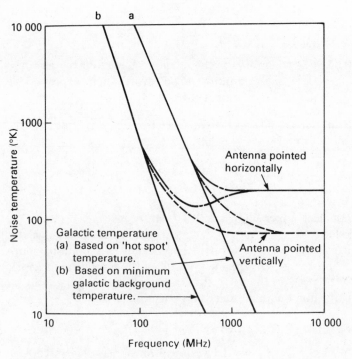

Fig. 2.12 Galactic noise temperature/frequency.

(b) Atmospheric noise: the effect on the signal transmitted through the atmosphere is not just one of attenuation but also it collects additional noise due to the fact that hydrometeors in the atmosphere act as a dipole antenna that is radiating a noisy signal. The atmospheric loss is proportional to the distance, the noise element is elevation dependent as defined in equation 2.10.

$$T_a = T_m(1 - \beta_o^{\,\cosec\,\alpha})\,^{\circ}\text{K} \tag{2.12}$$

where T_m = mean radiating temperature of the atmosphere

β_o = transmission coefficient of atmosphere through the zenith

$\cosec\,\alpha$ = angle of elevation of the antenna

(c) Earth radiation: the earth itself acts as radiator of noise which will be added into the receiving system. It is to some extent elevation dependent.

(d) Satellite/terrestrial interference: there is always a problem of interference between satellite and terrestrial services where the frequency bands coincide. The allowable level of interference is set by international regulation, the level is dependent upon the type of service, modulation methods, earth station characteristics etc. The earth station antenna sidelobe characteristics are defined in order to restrict the amount of signal picked up from terrestrial and satellite sources that are off axis.

2.8 ENERGY DISPERSAL

The amount of interference between satellite and terrestrial services is of crucial importance in the design of the overall system and its constituent parts; this has led to the development of techniques that ensure that for multi-carrier and television transmissions the maximum spectral energy that the satellite receives is limited to a level that minimises the interference caused to other systems. This technique is known as energy dispersal and is applied to the transmitted signal during periods of light loading.

If system designers simply take the maximum levels of allowable inter-ference due to other satellites, atmospheric effects and terrestrial interference then they are going to arrive at a system design that is too conservative, overcompensated and worse still, will require system components that are of higher specification and therefore of higher cost than is strictly necessary. The system boundaries are set in terms of equivalent noise power injected into a single telephone channel for an earth station–satellite–earth station link, this power being defined as a noise level psophometrically weighted one minute mean power for 20% of any month; this power should not exceed 1000 pWOp in total, in which no element of interference should exceed 400 pWOp. These limits are relevant for C Band and Ku Band satellite systems and depend to some extent upon the particular system design. In FM/FDM systems, which use frequency modulation, then the permissible levels of interference are related to the mean power level, which in turn is dependent upon the number of channels, rms deviation, channel bandwidth and the carrier frequency of the interfering signals. The protection ratio in this case is defined by:

$$Pr = 10 \log \left[\frac{P_w}{P_u} \right] + B \qquad (2.13)$$

where P_w = wanted signal power

P_u = unwanted interfering power

B = interference reduction factor

The factor B can be defined as:

$$B = \frac{P}{B_{ch}\left(\frac{f_m}{f_r}\right)^2 \times G(f)(f_d + f_m) + G(f)(f_d - f_m)} \qquad (2.14)$$

where B_{ch} = channel bandwidth in Hz

F_r = test tone rms frequency deviation in Hz

F_m = maximum baseband frequency in Hz

F_d = carrier frequency difference

P = pre-emphasis in the top channel

G = convolution of the normalised power spectrum of each system

The above relationship shows that as the modulation index increases the interference reduction factor is increased thus the baseband channel noise is reduced. This in itself allows closer satellite spacing but reduces the effective usable satellite channel capacity, for a specified limit of interference.

Chapter 3
Signal Processing

3.1 INTRODUCTION

The basic communication system would ideally transmit the information through the transmission medium, be it copper wire, optical fibre or the atmosphere, without loss or distortion and the receiver would produce the original information as input to the transmitter. This is evidently impossible as all three elements of the transmission system affect the original information.

The practical realisation of a transmission system is a trade-off between three parameters that are fundamental elements in communications:

- Information capacity in bits per second;
- Bandwidth in hertz;
- Signal to noise power ratio, S/N.

These elements are related in Shannon's equation:

$$C = B \times \log_2 \left[1 + \frac{S}{N} \right] \text{ bits/second} \tag{3.1}$$

This equation indicates that to achieve an increase in information capacity, either the bandwidth or transmitted power must be increased or the system signal to noise must be reduced. In practice a finite limit is set to the signal to noise ratio i.e. a system has a defined S/N for analogue and bit error rate for digital performance. Transmission bandwidth is a finite resource and therefore any transmission system must aim for the maximum utilisation of the bandwidth available i.e. the maximum information capacity per unit of bandwidth must be passed by the system.

There are however some circumstances where the signal/noise ratio of the transmission is the most important parameter, so that the bandwidth has to be increased in order to achieve the required information transmission over a particular medium, i.e. the medium of transmission cannot be altered or avoided and therefore must be overcome.

The history of communications is in many ways the history of the search for improved efficiency and the quality of signal transmissions. The methods used in this search are detailed in the following sections.

3.2   TYPES OF MODULATION

3.2.1   Introduction

In general the majority of communication systems transmit relatively low frequencies or bit rates. These signals do not propagate down a wire or through space without severe attenuation. However a propagation medium will often allow higher frequencies/bit rates to pass with less attenuation; this means that the transmission of information becomes possible if the speech/data is used to modify or modulate a higher frequency which acts as a carrier for the lower frequencies. This process is known as modulation.

A modulation scheme can take a variety of forms, which will be detailed in the following sections, where the frequency, phase or coding of the signal is varied to achieve the required transmission. The method of modulation is chosen for a variety of reasons:

Simplicity
The communication system must always be as simple as possible to implement, in order to reduce not only the cost of modem equipment but also the total cost of ownership which includes hardware costs and the cost of maintaining the equipment. Account must, therefore, be taken of the complexity, both for ease of maintenance and the training costs incurred to ensure that staff have sufficient knowledge to prevent loss of traffic and therefore loss of revenue. The choices open to the designer must therefore be judged on a cost effective basis even though the fundamental system performance must be met.

Bandwidth utilisation
As stated already the spectrum is a finite resource and the maximum utilisation of that resource is a major factor in any system design and one of the key determining facts in the modulation system to be used, as again bandwidth is money.

Signal/noise improvement
The designer must ensure that the modulation system used will provide a S/N improvement and an increased resistance to interference from outside signal sources. This is essential not only in terms of reducing the size of transmitters, for instance, but also to ensure that the communications will work under poor environmental conditions without error and therefore at the most cost-effective level.

3.2.2 Frequency modulation

Basic system

The first type of transmission used in satellite communications was that commonly used in terrestrial microwave systems i.e. frequency modulation. Frequency modulation was adopted because it had been found to provide an improved signal to noise ratio and it operates on the principle that the frequency of the carrier wave varies as the modulating baseband frequency changes. The effect of modulation is to produce a waveform that can be defined by equation 3.2.

$$\omega_m = A \sin \left(\omega_{ot} - \frac{d\omega}{\omega_a} - \cos \omega_{at} \right) \tag{3.2}$$

where ω_o = carrier frequency

ω_a = modulating frequency

ω = maximum frequency deviation from ω_o

Equation 3.2 shows that the frequency modulated envelope is of constant amplitude A, but having an instantaneous frequency variation. The modulation index of the signal is defined as:

$$M = \frac{d\omega}{\omega_a} \tag{3.3}$$

Equation 3.2 can be developed to show the sideband structure of the frequency modulated waveform, with each sideband being defined in terms of an amplitude calculated from Bessel function tables and a frequency spacing at a frequency of F_a from the carrier frequency. A typical FM sideband structure is shown in Fig. 3.1.

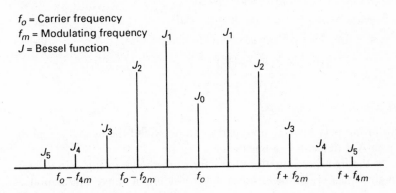

Fig. 3.1 Typical FM frequency spectrum.

The bandwidth requirements of a frequency modulated signal are greater than an AM modulated signal and it is fortunate that the sideband amplitudes fall away quite rapidly. In a practical system the designer neglects those sidebands that contribute less than 1% of the total energy; this bandwidth can be defined by Carson's Rule set out in equation 3.4.

$$B_{rf} = 2[dF + f_m] \tag{3.4}$$

where dF = multi-channel peak deviation in Hz

f = mid-frequency of baseband channel in Hz

The relationship between FM and AM noise performance is given by equation 3.5.

$$\frac{N_f}{N_{am}} = 20 \log\left(\frac{F_b}{F_{dm}}\right) - 4.72 \text{ dB} \tag{3.5}$$

where N_f = FM baseband noise

N_{am} = AM baseband noise

F_b = audio baseband frequency

F_{dm} = maximum FM deviation

The improvement is at the expense of bandwidth and in practical frequency division multiplex systems it is necessary to restrict the peak to rms deviation ratio in line with the equation 3.6.

$$dF = f_r \times G \times L \tag{3.6}$$

where f_r = rms test tone deviation/channel in Hz

G = peak/rms factor (usually 10 dB)

L = multi-channel loading factor in dB

$= -1 + 4 \log N [N = 12\text{--}240 \text{ channels}]$

$= -15 + 10 \log N [N > 240]$

FM/AM improvement threshold

The relationship between the input carrier to noise and the output signal to noise for an AM detector is a linear one.

However the FM discriminator characteristics, shown in Fig. 3.2, do not maintain this linear relationship over the whole input range but exhibit a threshold effect, at low input signal to noise ratios, where the noise ratio advantage compared to AM is seriously eroded. The performance of an FM discriminator is defined in terms of its threshold point, a point which is defined as the input signal to noise ratio value where the output signal to noise is 1 dB relative to the linear extension as shown in Fig. 3.2 and is approximately 10 dB for a conventional discriminator.

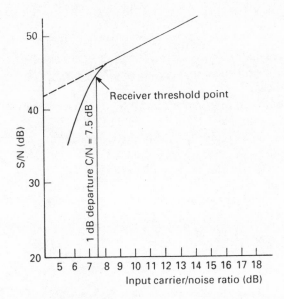

Fig. 3.2 FM threshold characteristics.

The FM improvement is due to the fact that the demodulator presents a reduced bandwidth to the signal and compresses it into its original baseband frequency limits. In addition to the basic characteristics of the discriminator the system will use pre-emphasis, which takes account of the triangular noise spectrum, associated with FM, to emphasise the higher frequencies for transmission to give a gain in effective signal to noise.

The threshold point of a FM detector can be extended to give an improvement in system performance and this is dealt with in chapter 4 on earth station systems.

3.2.3 Phase modulation system

The use of phase modulation is mainly associated with digital transmissions, with a considerable amount of development being put into phase shift keyed (PSK) systems in order to pass more information through the same bandwidth.

The simplest PSK system is binary PSK, which is a coherent system requiring a carrier recovery system that is unambiguous with regard to carrier phase. In this case the transmitted signal can be represented by equation 3.7.

$$\omega_t = A(\cos \omega t + \theta_t) \tag{3.7}$$

where f = carrier-frequency

$\theta_t = 0°$ or $180°$

The signal has two different states; $A \cos \omega t$ and $A \cos[\omega t + \theta]$ where θ is 180°. The BPSK modulation is equivalent to a digitally implemented double sideband suppressed carrier AM.

The carrier recovery system used in practice for BPSK introduces phase ambiguity into the recovered signal. This ambiguity would be 180° and would introduce 100% errors, but it can be removed by differentially encoding and decoding the transmitted/received signal, which removes the need for a carrier recovery system. The spectral efficiency of phase shift keyed systems is measured by the number of bits that can be transmitted in one hertz of bandwidth and is defined in equation 3.8.

$$S_d = 2A^2 \times T_b \left(\frac{\sin \pi \times f \times T_b}{\pi \times f \times T_b} \right)^2 \tag{3.8}$$

where T_b = bit duration

f = data rate

The spectral efficiency can be improved to 1.5 bits/Hz by the use of four-phase PSK (QPSK), wh.ch is probably the most popular PSK transmission method, combining as it does reasonable spectral efficiency and simple modulation and demodulation circuitry. The QPSK system can be regarded as two binary systems operating in quadrature with stable phase states as shown in Fig. 3.3.

3.2.4 Pulse code modulation (PCM)

Introduction

The improvement in microcircuit technology in the form of integrated circuits, LSI and VLSI, has meant that digital processing of signals has become not only a simpler matter but also more cost effective, enabling the use of PCM in digital communications systems.

The principle of pulse code modulation is derived from the application of sampling theory which states that:

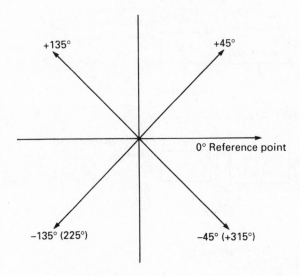

Fig. 3.3 QPSK phase states.

'Any continuous signal, which has a maximum frequency of f_m can be constructed from a set of discrete samples if the interval between the samples is less than half the maximum frequency or expressed another way the sampling frequency must be greater than $2f_m$.'

The process of producing a PCM signal happens in a number of phases:

(a) The signal to be transmitted is sampled to produce a signal that consists of a series of narrow pulses at sampling frequency f_s and is essentially a pulse amplitude modulation (PAM).

(b) The PAM signal is then quantised i.e. each pulse is given an amplitude value.

(c) The PAM quantised signal is then encoded by converting the quantised level into a digital code.

In order to implement a PCM system it is necessary to determine:

● The optimum sampling rate;
● The coding/encoding method to be used.

The above operations define a typical PCM implementation as shown in Fig. 3.4.

The input signal is applied to the sampling circuit via a low pass filter, which defines the upper boundary of the input frequency, which should be the design frequency f_m. This audio signal is sampled at a frequency f_s, which

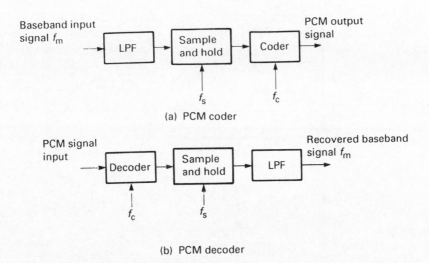

(a) PCM coder

(b) PCM decoder

Fig. 3.4 PCM system implementation.

pulse amplitude modulates the signal. The samples are then applied to a coder circuit which quantises each pulse to the quantisation levels set in the design and the coder output consists of a binary pulse train, each code word being made up of *n* bits. The clock rate that the coder must use is determined by the sampling frequency and the number of bits in each code word. Thus:

$$f_c = f_s \times n \tag{3.9}$$

The PCM system bandwidth is given by equation 3.10.

$$B_{pcm} = \frac{n \times f_s}{2} \tag{3.10}$$

To reverse the process the system is reversed and a PCM signal is applied to a decoder that uses a clock signal for the decoding and again sampled at f_s, before being filtered to recover the audio signal.

The quantisation level and the number of quantisation steps is a function of circuit complexity and sensitivity, coupled with the fact that the quantisation process introduces an error in as much that the actual sample level can be half a quantisation step away from a particular operating position. The quantisation steps should, therefore, be as small as possible, in order to reduce this error, but this process of reduction will be limited by the circuit's ability to differentiate between each quantisation level. The quantisation error voltage manifests itself as a sawtooth waveform whose amplitude is the quantisation step amplitude and its frequency is $2f_s$, assuming equal quantisation intervals. The quantisation dist٢ ٠tion thus appears in the form of a noise and the signal to noise ratio of the system is defined by equation 3.11.

$$S_{nq} = 10 \log[(s)^2 - 1] \, dB \qquad (3.11)$$

where s = number of quantisation levels

For standard values of s, say 256, the signal-quantisation noise ratio is given by:

$$S_{nq} = 10 \log s^2 \, dB \qquad (3.12)$$

Table 3.1 shows the effect of changing s on signal to noise.

Table 3.1   Effect of number of quantisation on signal/noise

Value of s	S/N	Code word bit no
32	30.1	5
64	36.1	6
96	39.6	6.6
128	42.1	7
256	48.2	8
385	51.7	8.6
512	54.2	9

The number of bits in the code word is dependent upon the number of samples and increases with the increase in the sampling rate and Table 3.1 shows that the number of quantisation intervals must be a binary function to give integers of n.

The S/N figures given in Table 3.1 are theoretical and will not be achieved in practice at low levels of input signal. One method of improving the performance is to compand the quantisation characteristic, as with pre-emphasis in analogue systems, to give smaller quantisation intervals and this characteristic is shown in Fig. 3.5 and is known as an A Law, which gives an effective compander gain.

The compander gain is defined by equation 3.13.

$$G_{sq} = 20 \log \frac{N_{si}}{N_{ss}} \qquad (3.13)$$

where N_{si} = number of step intervals

   N_{ss} = number of standard steps

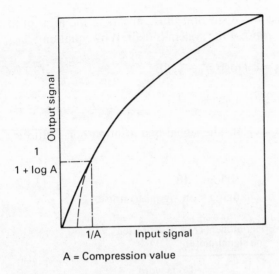

A = Compression value

Fig. 3.5 A law compression characteristic.

From the above equation, for 256 steps, this gain would be 24.1 dB.

In practice there are two standards that apply to PCM systems. The North American system devised by Bell is known as T1, which is defined in CCITT Rec 733. The system has 24 voice channels, using μ Law compression, contained within a frame that is 193 bits long. The 193 bits frame is constructed from 24 channels that contains 7 speech bits and 1 signalling bit, giving 192 bits per frame; the 193rd bit, at the end of the frame, is used for synchronising purposes. The T1 system uses an 8000 Hz sampling rate and has an overall rate of 1544 kbits. The use of a 7-bit coding produces a basic channel data rate of 56 kbits.

The European system, sometimes known as the CEPT system, is defined in CCITT Rec 732. The system uses A Law compression and has 32 time slots, comprising 30 voice channels, one synchronisation and alarm channel and one signalling channel. The frame length is 8 bits long giving a multiframe length of 256 bits. The system uses an 8000 Hz sampling rate, an 8-bit time slot giving a basic time slot of 64 kbits and a multi-frame frequency of 2048 kbits.

The basic speech channel is 56 or 64 bits, depending upon the standard used. This is built up in a hierarchy of channel sizes as shown in Table 3.2.

There is also a hierarchy of transmission data rates as detailed in Table 3.3.

The differences in the two systems mean that the interface between satellite and terrestrial systems requires specific equipment such as transmultiplexers.

Table 3.2   PCM standards

Hierarchy level	North America	Standard	
		Europe	Japan
1	24	30	30
2	96	96	120
3	672	480	480
4	4 032	1 440	1 920
5	–	5 760	7 680–115 200

Table 3.3   Transmission data rate hierarchy

Hierarchy level	North America	Transmission rates (kbits)	
		Europe	Japan
1	1 544	2 048	1 544
2	6 312	8 448	6 312
3	44 736	34 368	32 064
4	274 176	139 264	97 928
5	–	560 840	396 200

Differential PCM

The designer of communication systems is always looking to optimise bandwidth utilisation. There is one area of research that has been extensively investigated and that is information compression. The design boundaries are set by the quality of the communication and, provided this parameter is met, the signal processing that takes place before transmission should remove all unnecessary information. In sampling systems this is done by taking only an instantaneous view of the information. If the sampling system could predict what the successive samples would be then less information need be sent each time, in fact only differences need be sent.

If we consider PCM then the difference between successive samples is likely to be small, therefore the final output is to some extent predictable, especially if A Law quantisation is utilised. The predicted signal can either be based on the input signal or be obtained from the output signal.

A block diagram of the basic system is shown in Fig. 3.6, which uses output signal related prediction as this gives less errors than the simpler input signal prediction method. It can be seen that the transmit system uses the same signal recovery circuitry as the receive system producing identical Tx or Rx predicted values.

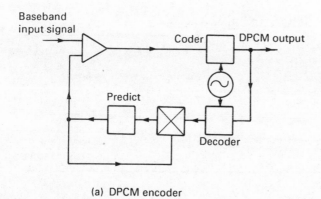

Baseband input signal

Coder

DPCM output

Predict

Decoder

(a) DPCM encoder

Decoder Multiplier

Recovered baseband output

DPCM input signal

Predict

(b) DPCM decoder

Fig. 3.6 Differential PCM system.

Delta modulation

A special form of differential PCM or DPCM is known as 'delta modulation', in which the predicted signal level is assumed to be the same as the last one that was processed. The prediction circuit then has to simply delay the signal by one sample period to obtain a prediction signal. The simplicity of the system means that it also has drawbacks and in the case of simple delta modulation the problem is that because of the rate of change of input signal the circuit cannot follow the input signal and therefore the system signal to noise becomes dependent upon the input signal characteristics.

The problem described above is known as 'slope clipping' and can be overcome by introducing a differentiator after the demodulator and by increasing the sampling rate or by increasing the quantum step size, which will increase the sampling error. In addition the delta modulation requires one processing circuit per channel and cannot share the quantising circuitry as in standard PCM. This was a considerable difficulty before the advent of VLSI but is now much cheaper and has raised the level of interest in this form of modulation.

There is a further development to delta modulation which can give improved performance; this is known as adaptive delta modulation. Here the quantisation step size does not remain constant. For a slowly changing input signal the step size is small, giving good S/N performance. For rapidly changing signals the step size increases; this counteracts slope clipping and again gives good S/N performance. Such a change of step size can be given any characteristics the designer requires to meet the system performance.

3.3 TRANSMISSION MODES

3.3.1 Introduction

The signal transmitted to and from the satellite consists of a baseband signal modulated onto a microwave signal. The transmission format depends upon the type of modulation being used and the method of organising the individual channels. The following sections delineate some of the possible methods.

3.3.2 Frequency division multiplex (FDM) systems

The basic analogue baseband signal for a single telephony channel is contained within a frequency band that extends from 300 Hz to 4000 Hz.

The individual channels are formed together by modulating the telephone baseband signal on to a carrier as shown in Fig. 3.7.

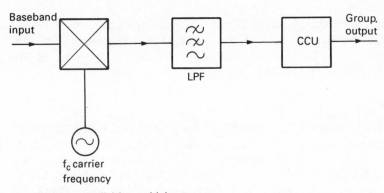

Fig. 3.7 Frequency division multiplex system.

The form of modulation is a single sideband suppressed carrier where the lower sideband is used and this is selected by a filter before being combined in a channel combining unit (CCU). The CCU first adds the lower sidebands, at 4 kHz intervals, to form a 12-channel group which occupies 48 kHz of bandwidth over the range 60–108 kHz; this is known as a basic group.

The basic group represents a low capacity system and for larger systems group translation equipment (GTE) is used to create larger groups, the first of which is a Supergroup that is made up of 5 groups, giving 60 channels over a frequency band of 312–552 kHz. This can be further expanded to form a Mastergroup, which consists of 5 Supergroups, comprising 300 channels over a frequency band of 812–2044 kHz. This is sufficient for analogue satellite systems which are specified up to 972 channels for INTELSAT V and VI series of satellites. In this case the INTELSAT Document IESS–300 details the frequency bandwidth and number of channels that can be used with that bandwidth as shown in Table 3.4.

Table 3.4   Number of channels in satellite bandwidth

Bandwidth (MHz)	Number of channels
1.25	12
2.5	24–60
5.0	60–132
7.5	96–192
10.0	132–252
13.0	252–432
17.5	312–432
20.0	432–552
25.0	432–612
36.0	612–972

The basic structure of the FDM system is set by CCITT & CCIR recommendations, which are in almost universal use. There are other standards in use particularly in the USA where the Bell Telephone System generated its own recommendations known as Bell System Practices (BSP).

FDM/FM

The FDM/FM system uses frequency modulation as its carrier modulation method and has been in use since the early days of satellite communications, with either single or multiple destination carriers being used. In the satellite the need to amplify multiple carriers, situated contiguous to one another in the frequency spectrum, means that TWT amplifiers are used to obtain a linear amplification and so minimise intermodulation distortion. If klystron amplifiers are used then special precautions have to be taken.

The major impairments present in the FDM/FM system are dependent upon the bandwidth of the satellite transponder and the number of carriers contained within that bandwidth. In chapter 4, which deals with earth stations, consideration is given to such distortions as intermodulation, intelligible cross talk and the effect of amplitude slope and group delay upon these impairments.

The other effects that must be considered are distortions, impairments and noise resulting from the following factors.

Interference between carriers in adjacent carrier groups

This interference is kept to a minimum by the use of guard bands to separate the groups and a rapid increase of filter attenuation out of band to reduce the signal available for causing interference. This guard band is a compromise between minimal interference levels and maximum available bandwidth and the necessity for it produces a transponder utilisation factor of approximately 90%.

Cross-polarisation effects

For frequency re-use systems, which use different polarisations to transmit and receive signals that are at the same frequency, there will be transfer of signal between polarisations due to cross-polarisation effects both in the transmission medium and the antenna sub-systems.

This effect can be minimised by ensuring that cross polarisation of the antennas is designed to be at least 35 dB. The transmission depolarisation effects are frequency dependent and have an increasing effect at higher frequencies. It is not possible to deviate from the internationally agreed satellite frequency bands, the use of which is also dependent on the frequency band utilisation.

Adjacent transponder interference

The interference between carriers is a function of frequency operation, controlled by filtering. If the transponder is fully utilised then those channels in use at the edge of the adjacent transponders will interfere with one another. The interfering signal is transmitted at the same time as the correct transponder output and received at the earth station, via a different path, which gives the effect of delay distortion due to the signals being different phases.

Earth station intermodulation

This is considered in chapter 4 and in practice a noise allowance of 500 pWOp is set; this noise limitation means that the power transmitted to the satellite has to be restricted to a maximum level of:

$$P_M = 26 - 0.06(\alpha - 10) \text{ dBW}/4 \text{ kHz} \qquad (3.14)$$

where α = satellite pointing angle.

Impulse noise

Impulsive noise is generated in the system by two major mechanisms if the carrier separation is small. Firstly bandwidth limitation: if an FM carrier is passed through a bandpass filter that is narrow compared to the occupied bandwidth and the carrier deviation is sufficient to take the frequency into the filter's attenuation slope, then again we have carrier attenuation. If this takes it below the noise then an impulse of noise will appear at the demodulator output.

The impulse noise requirements are controlled by the transmit and receive IF filters and these are designed with an elliptical function Tschebischeff characteristic with amplitude and group delay equalisation.

The FDM/FM system is used extensively in earth stations, operating in the INTELSAT system, due to the relative simplicity and cost-effectiveness of its hardware requirements. The major problems with the system are concerned with bandwidth utilisation efficiency and intermodulation distortion. The earth station transmitted power must be controlled to within specific limits in order to ensure that the up-link noise limitations are met; this places limitations on the earth station HPAs, which must be able to control its output to within tenths of a dB, as well as restricting the antenna gain tolerance which must take account of the changes caused by environmental effects.

The second effect is adjacent carrier interference: if a demodulator is operating close to the threshold and the noise level on the signal is greater than the wanted carrier at the same time as the noise carrier phase is approximately 180°, then impulsive noise will appear at the demodulator output; it is often called click noise.

3.3.3 Single channel per carrier/FM

The basic premise for the assigning of an analogue or data carrier to a single channel is to reduce the distortion introduced into the signal. The effect of signal impairment means that as the number of carriers accessing the satellite increases then the HPAs operating point must be backed off, leading to less power being available to maintain the channel capacity. This is more of a problem for low channel capacity systems and where the earth station does not need to sustain more than 60 channels. The INTELSAT system provides a single channel per carrier service that uses frequency modulation (SCPC/FM). The SCPC/FM transmissions actually modulate the individual voice channels on to a separate carrier. The INTELSAT organisation has developed its Standard B earth station so that it can operate with up to 792 channels with companding being used to double the capacity. A compander allows a larger number of channels to be accommodated on the same transponder by reducing the peak deviation of the FM signal by compressing the modulating amplitude. In addition to companding some systems use voice activation, which means that a carrier is transmitted only when the voice signal is present. This use of voice activation can more than double the number of channels available from the satellite, subject to bandwidth limitations.

3.3.4 Time division multiplex (TDM) systems

The TDM method is used in digital communication systems and is the major alternative to FDM/FM. The individual earth station transmits digital information, which may originally have been voice or data, to the satellite in bursts; each burst passes through the transponder and in essence occupies the total channel bandwidth on its own and therefore no intermodulation distortion can be produced and therefore the maximum transmission power can be used.

Time division multiple access (TDMA)

If TDM transmissions are organised so that a number of earth stations can access the same satellite transponder then we have one of the most important developments in satellite communication, leading to high capacity digital transmissions. The individual earth station bursts are transmitted at a specific time to its allocated time slot. The downlinked signal contains all of the individual signal bursts that have been transmitted to the satellite, formed into a framed signal that contains voice, data and video information, reference bursts and system control data. The earth station receives the total signal where the TDMA receiver selects that portion of the signal which is destined for it.

The TDMA system can provide greater available capacity, compared to FDMA, while at the same time giving greater system flexibility; this greater flexibility shows in that each individual earth station can change its capacity by modifying the burst formats, and also that the overall system is much easier to configure, in terms of its frequency plan, than FDMA, due to the fact that each station within the overall system has the same transmit and receive frequency. Additional capacity can be obtained in the TDMA system by use of digital speech interpolation (DSI) techniques. DSI is similar to voice activation in FDM/FM in utilising the fact that the speech activity in a normal conversation only occupies something like 40% of the total time involved. It is thus possible to obtain an effective transmission gain which is the inverse of the speech activity in any given channel; this gain could theoretically be as high as three times the standard capacity, but is usually conservatively estimated as two, giving an overall system improvement in capacity of three times over FDMA. In addition to the operational advantages TDMA becomes increasingly able to link in directly with terrestrial systems as they become digital in operation. There are disadvantages in the use of TDMA for those with low to medium capacity requirements would find the cost of implementing earth station equipment is considerably more than FDM/FM. For users below 750 channels it becomes economically difficult to justify, though this would depend upon traffic growth and the user's willingness to invest in earth station equipment greater than the initial requirements.

System implementation

The TDMA system requires three elements:

(a) A satellite with the necessary transponders;
(b) A reference station to synchronise the network of individual earth stations;
(c) Operator earth stations able not only to transmit and receive voice, data and video but capable of carrying out the necessary system operations under the control or direction of the reference station.

Reference station

The reference stations, operating within a TDMA system have, of necessity, to be totally reliable in that they organise and synchronise the activities of the network and the failure of the reference station would disrupt the communication capability of the network. It is therefore, necessary that the equipment used in the earth station must not only be highly reliable but must also utilise sub-system redundancy and in fact the whole station is duplicated, with a primary and secondary reference station in a hot standby configuration. In major TDMA systems, such as INTELSAT, there are a number of reference stations controlling the separate Ocean Regions. Each satellite transmits on spot and hemispheric beams; each spot beam contains two reference stations which actually control both spot and hemi-beams giving E/W and W/E interconnections. Each operational terminal is assigned its own reference bursts for control purposes and is controlled by the reference station in its own region.

The overall control of the system is completed by the regional reference station being itself controlled by a reference terminal in the opposite region. It can be seen that the software requirements for the overall system are immense, as indeed is the need to keep the software at all reference stations updated to the same status.

The reference station has to be able to determine the satellite position in real time and is equipped with ranging equipment for this purpose. This ranging measurement is also made by traffic stations. The system timing is achieved by transmissions of a reference burst that defines the TDMA frame, multiframe and control superframe. The ranging measurements provide information which tells the operating station that it must vary its transmit timing and this is relayed in the reference burst. This timing variation of each operating station is also monitored by the reference station and for any error it transmits a command back for the timing to be corrected. It is also part of the reference station's function to detect burst failure and prevent the operating station from continuing to transmit if the failure persists for more than a pre-determined time.

System earth stations

The operating earth stations have two basic functions: first to process the information to be transmitted, converting it into bursts of data which are ordered into frames, encoded and modulated, usually with multiple four-phase PSK, and transmitted to the satellite. The second function is to receive the total signal transmitted to the satellite and separate out the data specific to that earth station. A typical basic frame is shown in Fig. 3.8.

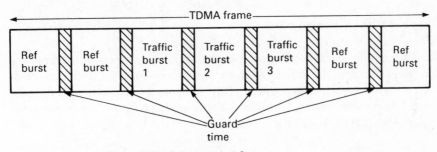

Note: CEPT Frame period 2 ms
Traffic burst 33.89 µs + preamble (176 bits)

Fig. 3.8 Typical TDMA frame.

The reference bursts are transmitted from two separate reference stations for reliability purposes. The traffic bursts contain not only traffic data but also control and order wire information. The individual bursts are separated by a guard time which is inserted to prevent interference between traffic bursts. In INTELSAT this is set at 48 symbols, with a framelength of 2 ms. The reference burst format is shown in Fig. 3.9.

176 bits	24 bits	8 bits	8 bits	32 bits	32 bits	8 bits
Carrier and bit timing recover	Unique word	Tele-type	Service channel	Voice order wire	Voice order wire	Acqui-sition and sync

←————————————288 bits————————————→

Fig. 3.9 TDMA reference burst format.

The earth station modems require a finite time to lock on to and synchronise with the incoming signal and therefore the reference burst begins with 176 bits that allow the synchronisation process to take place. The unique word differentiates the reference and traffic bursts and resolves the QPSK phase ambiguity. The frame also includes teletype and order wire carrying information – the final eight symbols carry information which controls the terminal's acquisition. The traffic burst is shown in Fig. 3.10.

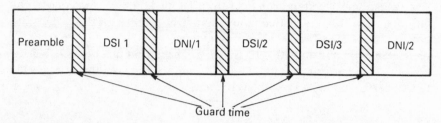

Fig. 3.10 TDMA traffic burst format.

Each traffic burst is referenced to the initial burst transmitted from the reference station; the individual traffic bursts are separated from each other by a digital speech interpolation (DSI) assignment channel, with the sub-burst being in either DSI units containing voice traffic or non-interpolated digital units (DNI), which accommodate data and non-interpolated traffic.

Frame efficiency

It can be seen from the above that actual traffic bursts require a certain amount of overhead. Transmission efficiency is defined as the ratio of useful information transmitted to the total information transmitted. Thus:

$$F_e = \frac{mS_b}{m(S_b + S_g)} + S_p \tag{3.15}$$

where m = number of traffic bursts

S_b = number of symbols per burst

S_g = number of symbols per guard time

S_p = number of symbols in the preamble

The number of bits of information is dependent upon the modulation used, but for QPSK each symbol contains two bits of information; the number of bits in a traffic burst is dependent on the data rate and the length of the traffic burst.

System efficiency

The system efficiency is not just a function of the frame efficiency but has to take into account the number of terminals.

$$S_e = T_f - [T_R + nT_g + n(n-1)] \times \frac{T_a}{T_f} \qquad (3.16)$$

where T_f = frame time

T_R = ranging time

T_g = guard time between transmission

T_a = number of terminals in the system

The efficiency of the transmission can be increased by the use of longer frame lengths as the various preambles and guard times should remain the same. There is a premium to pay in terms of cost due to the larger memory needed, as this is proportional to frame time; an increase in the phase noise; and if the frame period goes beyond 125 microseconds, the analogue storage is difficult to use.

Satellite switched TDMA

There is another form of TDMA that has an advantage over normal TDMA which is satellite switched system SS/TDMA. It is necessary for the overall system to use multi-spot beam operation, with the satellite having independent transmit and receive systems, with very high speed circuitry, able to connect between Tx and Rx systems. The advantages of SS/TDMA arise from the fact that communication can be achieved between different regions using spot beams and therefore higher power is available and the effective bandwidth can increase in proportion to the number of beams.

The extra complexity of both satellite and earth station provides an additional cost in equipment; the satellite must switch rapidly to avoid reducing system efficiency and the earth station must have additional equipment that will allow it to automatically hop between transponders, ensuring that each receive chain is equalised in absolute gain and phase.

3.4 CODING METHODS

Coding of a signal, whether analogue or digital, is carried out for specific reasons, all of which are an attempt to extend the communication capability of the existing link. The link has to achieve a particular performance quality, related to signal to noise or bit error rate; it has to meet the performance for a particular transmitted power or within a particular bandwidth; in many cases the link performance cannot be met without coding.

In coding a signal the basic information is modified in such a way that the transmitted information contains more than the orginal message but is able to achieve a reliable communication under circumstances which would normally not allow it. In practice the use of coding is similar to modulation and is applied for the same reasons. The coded signal is modified by additional bits as shown in Fig. 3.11.

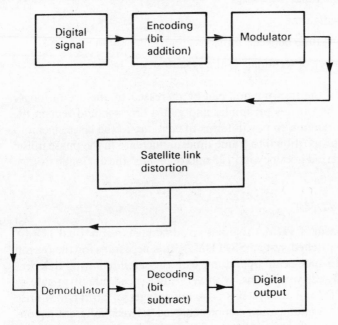

Fig. 3.11 Coding employed in satellite communications.

The signal presented to the demodulator in a satellite communication system differs from the original signal by the addition of:

(a)   Additional coding bits;
(b)   Intermodulation distortion due to the HPA;
(c)   Cross polarisation distortion;
(d)   External noise.

The demodulator has to differentiate between the actual digital signal and the received signal which is distorted by the degrading elements. The demodulator can be either of two forms:

(a)   Soft decision detection;
(b)   Hard decision detection.

Hard decision detection is a simple decision that in practical terms results in an appropriate reconstruction of the original signal. The information so recovered is specific and contains little information to assist in the decoding process. Soft decision detection does not produce the transmitted signal from one operation on the received signal but takes a number of decisions which provide partial detection, but at the same time produces further information that assists the detection process, before coming to a final conclusion.

The coding process is not only concerned with ensuring that the original information can be transmitted through a practical system but that it can also provide a measure of error correction which, although bandwidth-inefficient, gives improved performance in satellite communication systems that are not bandwidth limited but have the problem of transmission delay.

A major type of coding system is known as block coding; in this method the input signal to the encoder is divided into a block of digits, the number of digits being pre-defined. The output from the encoder is also divided into blocks but not of the same number of digits. The block code is simple to implement as only one block is considered at any one time.

The simple block code is best used when the information to be processed is of that form. For more complex signal structures the code must be more complex and instead of single independent blocks of digits the input becomes a number of blocks of digits coded into a single block in a linear fashion, ensuring simplicity of decoding. The generic name for this multi-block code is a tree or convolutional code. In convolutional codes the code is defined as having a particular rate, an example being the INTELSAT Intermediate Data Rate (IDR) system, which uses $\frac{1}{2}$ or $\frac{3}{4}$ Rate codes. The code rate is determined by:

$$R = \frac{k}{n} \qquad (3.17)$$

where k = the number of blocks decoded

   n = the number of digits in a block

The higher the code rate the less the number of redundant, i.e. non-information carrying digits, but the less powerful the code in terms of error correction. The error correcting power of the code is increased if the number of blocks stored in memory is increased and here the system will be efficient for long transmissions but have too high an overhead for short length messages.

One of the fundamental codes, developed for error correction systems, is known as the Hamming Code. This code can correct the error of any bit in a block of bits. The additional bits added to a block are inserted at all positions of 2 to the power n in the data block i.e. 1, 2, 4, 8, 16 etc. The other parameter of importance that must be defined is the Hamming distance, which is defined as:

$$d = 2t - 1 \qquad (3.18)$$

where t = number of errors corrected

The designer may not wish to correct all the errors in a signal, as the system can tolerate a certain amount of error, but may wish to know the true system performance and so must detect all the bit errors, in order to trigger alarm systems. In this case the forward error correction (FEC) must detect errors as defined in equation 3.19.

$$e = d - 1 \qquad (3.19)$$

The above applies to block codes for random error systems but can be applied in a general form to convolutional codes, in which the Hamming distance is between blocks rather than block words.

The decoding process is obviously more complex when FEC is applied to a signal and even for a small number of data blocks k, then the decoder requires a complex algorithm. The development that has taken place to simplify the decoding process is the use of another form of coding, viz cyclic and Read–Solomon; cyclic codes are part of a family of codes known as BCH, an acronym for Bose, Chauduri and Hocquenghem, the names of the codes' inventors. A cyclic code is produced by producing a new code word for every cyclic shift of a codeword. The advantage of BCH codes is that they not only produce improved system performance but are easier to decode than general cyclic codes. The Read–Solomon or RS codes are long cyclic codes that have many advantages over convolutional codes in their ability to process signals that are subject to burst errors, to deal with short messages and they also require simpler circuitry to implement.

The most familiar form of coding, used in satellite communication systems where error correction systems are most required, is the one that uses the Viterbi algorithm to decode the received signal, particularly with satellite operation using PSK modulation on channels that are power limited. The significance of the power limitation is that the satellite can either be power or bandwidth limited; if it becomes bandwidth limited then coding only exacerbates the problems because it increases the bandwidth required.

The Viterbi method is based on the fact that whereas it is possible to determine the probability of correct code symbol sequences on all of the possible paths through a trellis, it is not necessary to do this if limitations are placed on the trellis code structure. If the decoder assessed all paths then the decoder would have to perform the number of operations defined in equation 3.20.

$$n = 2^L \qquad (3.20)$$

where L = number of bits in an information sequence

The Viterbi decoder however requires the number of operations as defined in equation 3.21.

$$n = L \times 2^{k(K-1)}$$

(3.21)

where k = number of paths

K = trellis depth

3.4.1   Coding gain

The definition of coding gain becomes difficult due to the effect of the coding process itself; the coding system alters the basic information, i.e. it adds redundant information to the signal, therefore although the system can have improved performance in terms of bit error rate it is less efficient in terms of bandwidth or power required to transmit the extra digits.

This effect produced by the coding system means that the code designer has to ensure that the improvement in the operating point of the system in terms of the signal to noise ratio must be greater than the power increase, in dB, needed to transmit the extra digits. The coding gain is given in equation 3.22.

$$G = 10 \log R \times d$$

(3.22)

where R = code rate

d = convolutional code distance

A further difficulty is that in a practical system, which must remain in synchronisation (otherwise re-synchronisation must take place with an increase in errors in transmission), the demodulator must receive a certain quality of signal. Coding reduces that quality and therefore the demodulator input signal must have a higher signal input to noise ratio. For a system that has a coding rate of R the demodulator input penalty is:

$$L_c = 10 \log \frac{1}{R} \text{ dB}$$

(3.23)

Thus for the IDR code the coding process must produce 1.24 dB of coding gain before any real improvement is realised.

3.4.2 Spread spectrum coding

The access method is known as code division multiple access (CDMA) or spread spectrum multiple access (SSMA). The classic definition of a code division system is that it is one in which the average energy in the transmitted signal is spread over a much larger bandwidth than the required information bandwidth.

The characteristics of a spread spectrum signal gives it major advantages compared to other systems, in that it can provide reliable communications, even in the presence of high level interfering signals. It is not easy for an outsider to listen in to the transmission, first because the spreading process makes the transmitted signal resemble noise and thus not perceivable without special receivers; and secondly the code division processing ensures that only a receiver that is programmed with the same code as the transmitter can receive the message. This capability means that multiple-user systems, such as VSAT networks, can have all the users transmitting in the same frequency band, without high interference between transmissions, provided the codes used have a low cross-correlation function. In addition to the above characteristics the spread spectrum system provides a very high reliability of transmission because it is less affected by fading and multi-path effects, an effect that gives a greater improvement as the ratio of transmit bandwidth/information bandwidth increases.

The signal/interference performance of SS systems is defined for two particular situations; first where the interfering signal bandwidth is greater than the wanted signal bandwidth and secondly where the interfering signal bandwidth is less than the signal bandwidth. The two equations defining the signal/interference performance are:

For $B_i \leqslant B_s$

$$S_I = 2B_s \times T_s \left[\frac{A}{\sin A} \right] \qquad (3.24)$$

For $B_i \geqslant B_s$

$$S_I = \frac{2[B_i \times T_s]}{\frac{\sin B}{B}} \qquad (3.25)$$

where B_i = interfering signal bandwidth

B_s = wanted signal bandwidth

$S_I = \dfrac{\text{output signal/interference ratio}}{\text{input signal/interference ratio}}$

T_s = time duration of input signal information

$$A = \frac{\Delta\omega}{2B_s} \quad B = \frac{\Delta\omega}{2B_i}$$

ω = radian frequency difference between carrier of transmitted and interference signal.

There are three main forms of producing a spread spectrum signal.

Direct sequence spread spectrum (DSSS)

The basic system is shown in Fig. 3.12.

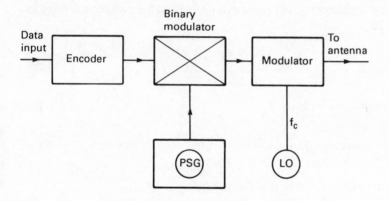

(a) Direct spread spectrum transmitter

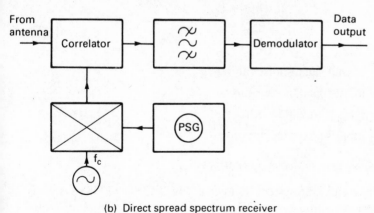

(b) Direct spread spectrum receiver

Fig. 3.12 Direct spread spectrum system.

The digital information to be transmitted is first encoded in a binary encoder. The output of this encoder is applied to a modulator and is modulated with a pseudo-random noise signal to give a spread spectrum signal.

This modulator output is then further modulated by a local oscillator carrier frequency; the final modulator output is then transmitted. The bandwidth of the pseudo-random signal is up to 1000 times that of the binary encoded signal.

The SS receiver consists of a correlator which multiplies the received signal with a pseudo-random signal coded signal and the correlator output is passed through a filter which has a bandwidth equal to the information bandwidth.

The system implementation produces a processing gain which is defined in equation 3.26.

$$10 \log A_p = \left[\frac{S}{I}\right]_{in} - \left[\frac{S}{I}\right]_{out} \tag{3.26}$$

where A_p = processing gain

$\left[\dfrac{S}{I}\right]_{in}$ = input signal to interference ratio

$\left[\dfrac{S}{I}\right]_{out}$ = output signal to interference ratio

The processing gain can also be defined in terms of signal to noise, RF bandwidth and signal time duration. This relationship is shown in equation 3.27.

$$A_p = \frac{S_o}{S_I} = 2\, B_{RF} \times T_s \tag{3.27}$$

where S_o = output signal-noise ratio

S_i = input signal/noise ratio

B_{RF} = RF signal bandwidth

T_s = input signal time duration

Frequency hopping spread spectrum (FHSS)

The basic form of an FHSS system is shown in Fig. 3.13. The FHSS system is basically a frequency shift keying (FSK) system which utilises a pseudo-noise sequency generator to shift the carrier frequency.

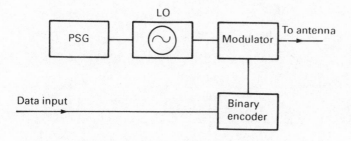

(a) Frequency hopping transmitter

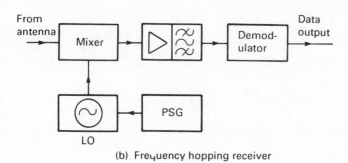

(b) Frequency hopping receiver

Fig. 3.13 Frequency hopping spread spectrum system.

The pseudo-noise sequence signal directly controls a frequency synthesiser, whose output frequency changes randomly, this synthesiser output is then modulated by encoded data before transmission. The transmitted signal is spread in spectrum, but differs from the direct system in that the instantaneous frequency spectrum consists of single frequency. The FHSS receiver takes in the signal from the satellite and mixes it with the pseudo-noise sequence; the mixer output is filtered before demodulated. The filtering decreases the output noise and gives the system a processing gain as defined in equation 3.28.

$$A_{pg} = \frac{B_{if}}{B_i} \tag{3.28}$$

where B_{if} = IF filter bandwidth
        B_i = information bandwidth

Non-sinusoidal carrier (NSC) spread spectrum

The NSC method of producing a spread spectrum signal is shown in Fig. 3.14

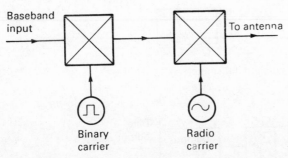

Fig. 3.14 Non-sinusoidal carrier (NSC) spread spectrum system.

The baseband signal is multiplied by a binary carrier signal that produces a spread signal; this spread signal is then multiplied with a radio carrier before transmission.

3.5 MULTIPLEXED ANALOGUE COMPONENT (MAC) SYSTEMS

3.5.1 Introduction

The terrestrial television services use an AM modulation system which is incompatible with the method of transmission proposed by the WARC Conference for the direct transmission of television by satellite (dbs). In the WARC proposals the preferred method would use FM modulation, as satellite distribution systems do at present. The problem that arises is twofold and based on both the use of FM modulation and the future possibility of high definition television (HDTV). The basic television signal has three major components:

- Chrominance information;
- Luminance information;
- Sound.

The above signal components are encompassed in a bandwidth of 4.4 MHz and in an AM system have a relatively flat frequency response over the whole of that bandwidth, giving similar signal to noise ratios for the chrominance and luminance signals. However for satellite broadcasting of television, where FM modulation would be used, there are three basic

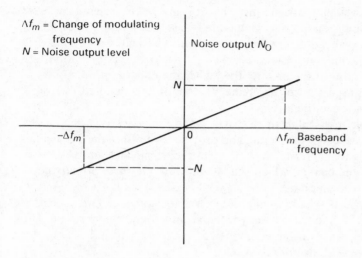

Fig. 3.15 FM noise characteristics.

problems. Firstly the noise characteristic of an FM signal is not flat, as shown by Fig. 3.15.

The FM effect is apparent when the chrominance signal is processed in that the high frequency, high level noise is demodulated down to appear in the baseband as a low frequency noise that is visible on the picture as a distortion.

Secondly the bandwidth of the signal is greatly increased in order to contain all the transmitted information.

Thirdly the FM threshold is affected, which causes a very rapid deterioration of signal to noise ratio below threshold as shown in Fig. 3.2 in section 3.2.2. The noise is transformed into video noise by de-emphasis network resulting in monochrome picture streaking. The need to find a method of improved transmission has also been pressured by the developments in high definition television, which will require considerably more bandwidth and thus lead to digital processing of the signals before transmission.

3.5.2   The development of (MAC) systems

Basic principles

The evolution of MAC systems began with the development work carried out by the Independent Broadcasting Authority in the UK and has subsequently been subjected to considerable research effort funded by the European Broadcasting Union (EBU), resulting in a variety of sub-sets of the MAC system each suitable for various types of transmission.

If we first consider the basic principles of any MAC system: the basic colour television system produces a picture by use of a scanning spot, which produces an image by scanning across the screen. The quality of this image is a function of the number of lines per picture, the picture presentation rate and the aspect ratio of the picture; the higher the number of lines, presentation rate and aspect ratio the greater the bandwidth required for transmission without loss of quality.

The colour picture is made up of two elements; luminance which contains the brightness aspects of the picture and provides contrast in the picture, i.e. a range of brightness from black to white; chrominance or the colour components of the picture, which can be analysed into three primary components, red, green and blue.

In practice the baseband signal comprises a luminance component, a colour burst on a sub-carrier, a sync burst and sound information. The MAC system takes the various components, separates them and places them in time sequence as shown in Fig. 3.16.

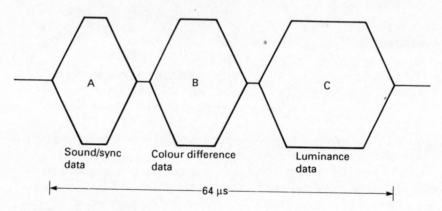

Fig. 3.16 Basic form of MAC signal.

The time sequenced signal consists of:

- Section A, which is a data burst that carries both sound and synchronisation information; this data burst lasts for 10 microseconds and has 206 bits per line, giving an instantaneous data rate of 20.25 MHz and a mean bit rate of approximately 3 Mbits. This signal will carry eight quality sound channels.
- Section B, which represents the colour difference signal components, with alternate lines containing U, Blue difference and V and Red difference components. The signal is processed so that each colour difference line is compressed in time by a factor of 3:1 thus occupying a time span of 17.5 microseconds.

- Section C, which contains the luminance information which is compressed by a factor of 1.5:1 to occupy a time period of 35 microseconds.

The residual 1.5 μs is used as a guard band between signal bursts.

The effect of the compression is to increase the baseband bandwidth so that the total required is 8.4 MHz, of which the luminance signal occupies 5.6 MHz. The above description refers to a C MAC system due to the form of the sound, known as Type C that is suitable for a European environment. The above characteristics will allow the C MAC signal to be transmitted with a 27 MHz bandwidth, which is that allowed by the WARC 77 recommendations. The MAC system has been developed to provide suitable variants for different applications and transmission requirements. The advantages of C MAC are:

(a) It is relatively immune to the non-linearities associated with FM;
(b) Zero cross-luminance distortion;
(c) Zero cross-chrominance distortion;
(d) Suitable for future HDTV transmissions;
(e) Compatible for scrambling systems.

B MAC systems

The luminance and chrominance signals are formatted in the same way as for C MAC, giving the same advantages. The difference in this system is in the data burst that carries the audio and data information; the data signal is at a rate of 1.86 Mbits and the burst can provide:

(a) Six high quality digital audio channels;
(b) Teletext;
(c) Message services;
(d) Control channels for service and subscriber descriptions.

The audio channels take advantage of the advances in modern technology and use a Dolby enhanced delta modulation scheme. Each channel uses:

(a) A bandwidth of 17 kHz;
(b) Dynamic range of 90 dB;
(c) Data rate of 251.7 kbits.

The channels are individually encrypted and are flexible enough to be configured as mono, stereo or four-channel systems.

The teletext service information is carried in the vertical blanking interval and uses two level symbols. The format is:

Line 1–8   Clock and sync recovery control data, conditional access and subscriber control.

Line 9–13   Teletext data consisting of 40 ASCII characters per line, i.e. 200 pages of text on a 625-line system.

Line 14–21   Additional lines for text to double the capacity.

The message service is carried in the teletext data and can be directly displayed or stored for later display in the teletext pages.

Encryption is an essential part of satellite TV, especially for subscription services where income is derived directly from the viewer. This is achieved by introducing pseudo-random delays between the active lines.

The B MAC system was developed by Scientific Atlanta and is in use in the Australian Satellite System for its HACBSS service to individual houses and isolated communities.

C MAC system for satellites

The basic picture format has already been described in the introduction to section 3.5.2. The sound and data information is transmitted in data packets that are 751 bits long, which for a 625-line system requires 162 packets and 4050 packets per second. The basic format was shown in Fig. 3.16. In this the sound/data packets are situated between the video information and usually take the form of an isolated data burst or sound channel. The sound and data signal is used to modulate an RF carrier and the modulated carrier is contained in the line blanking interval of the modulated video carrier. The time division multiplexing occurs at IF, with switching taking place between FM Video and digital sound/data, in order to maintain phase continuity. Each packet is identified by an identification code. The signal also provides for access control to allow for subscription TV systems to be operated. The total signal is scrambled; the picture scrambling can be accomplished by breaking the colour difference part of the MAC signal into two parts, as shown in Fig. 3.17. The two parts are then reversed; a process which can also be applied to the luminance signal, as shown, giving a very secure signal. The system is further randomised by making the cutting process operate from a pseudo-random number; the breaks introduced into the signal increase the spectral components and this must be restricted by filtering. The sound and data signals are also scrambled by adding a pseudo-random data stream.

C MAC systems for cable

The C MAC specification was developed for dbs operation and as a consequence is optimised for that application. In practical applications the dbs signals will be received in two ways; firstly as an individual signal into each home, via the owner's indoor receiver, and secondly at a cable head end for distribution to houses over a cable network. The ideal would be that the

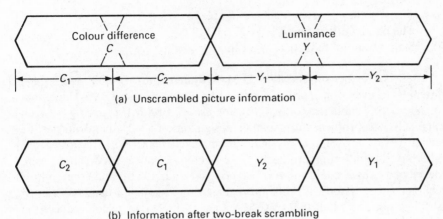

(a) Unscrambled picture information

(b) Information after two-break scrambling

Fig. 3.17 CMAC scrambling process.

format would be optimum for both cable and satellite but this is not possible for the following reasons:

(a) The dbs system uses FM and PSK modulation methods, while the cable system has to arrange a wide variety of services and is therefore more restrictive in bandwidth.

(b) If the C MAC system signal is translated down to a first IF, in the same band as the TVRO input signal, the system could handle up to 20 TV channels, assuming a 36 MHz bandwidth per channel, which allows for the 27 MHz bandwidth required for the video signal and the guard bands between channels. The majority of cable systems utilise coaxial cable, not optical fibre and this in itself restricts the number of channels as each cable in the tree and branch cable systems carry all the channels being distributed.

(c) The two systems have different carrier to noise ratios, with cable being much higher.

D MAC systems

The C MAC system is unsuitable for transmission on cable systems because of bandwidth conflicts. This has led to more research into further MAC processing systems, which again has concentrated on the sound/data requirements with the video processing remaining the same. The design aim was to reduce the overall bandwidth towards the theoretical bandwidth of 10.125 MHz, which would accommodate the sound data channel rate of 20.25 MHz, the bandwidth of the baseband binary singal from the demodulated PSK signal. The bandwidth requirements have led to the use of multi-level coding, in this case a duobinary system, which has the following characteristics:

(a) The three-level system which is simple to implement;
(b) The bandwidth required is 11.37 MHz;
(c) Signal to noise reduction of 3 dB.

There is a further sub-set of the D MAC system known as D2 MAC which attempts to overcome a problem remaining when one uses D MAC, i.e. that many existing cable networks in Europe use a 7 or 8 MHz spacing between channels. The problem with having a design aim of 5 MHz bandwidth is the actual realisation of the physical system, particularly with regard to the filtering. The filtering problems were overcome when SAW filters were developed, which could provide the necessary bandwidth and edge roll off characteristics. Even with this improvement the baseband bandwidth must still be reduced to half that of C MAC i.e. 10.125 MHz, which reduces the flexibility of the system as it allows only four sound channels to be carried.

In conclusion, Table 3.5 shows the detailed specification for each of the MAC types.

Table 3.5   MAC standards

Parameters	C MAC	D MAC	D2 MAC
Video	Identical format compatible for HDTV		
Data rate	20.25 Mbits	20.25 Mbits	10.125 Mbits
Packets/sec	4100	4100	2050
Audio	4/15 kHz	6/15 kHz	4/15 kHz
Modulation	QPSK	Duobinary	Duobinary

The D2 MAC Standard is being accepted as the transmission standard for domestic satellite television broadcasts. It is the most flexible of all the standards that have been developed and can be used not only for dbs and satellite television but is suitable for the transition to HDTV as that standard develops over the next few years, e.g. the EUREKA 95 Project due for completion in the early 1990s.

The overall system has very positive advantages over D MAC in that the receive system can have a smaller antenna because of the 2.5 dB signal advantage that it has. The system can also give a more flexible performance on data transmission because of its greater capacity and thus extend the revenue earning capacity of the dbs operators.

Chapter 4
Basic Earth Station Systems

4.1 INTRODUCTION

The basic form of any earth station is the same regardless of the system in which it is used. The elements of the system are shown in Fig. 4.1 and can be regarded as a two-way microwave communication link that requires certain specialised elements in order to operate in the particular environment associated with satellite communications.

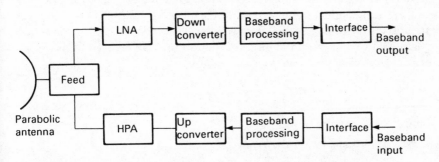

Fig. 4.1 Basic earth station form.

In this chapter each individual element of the earth station will be considered in detail but it is first necessary to consider the basic performance requirements of any earth station. The basic system parameters are usually set down as mandatory requirements for the specific satellite system and each earth station must conform to these standards before the user is allowed to access the satellite and so establish communications. This use of mandatory requirements allows the satellite system designer to take into account the various elements involved in calculating earth station performance and trading off the performance of each separate element involved in the earth station realisation, such as antenna, low noise amplifier (LNA) and high power amplifier (HPA).

The fundamental parameter in any earth station design is the figure of merit, G/T, which is the ratio of the system gain and the overall system noise temperature, where the system gain is the effective antenna gain, taking into account the actual antenna gain and feed losses. The system noise temperature is the addition of the effective receiver input noise and the total antenna noise contribution. The front end of any earth station has to provide signal conditioning that will allow the recovered baseband signals to have a specific carrier to noise (C/N) for particular forms of transmission such as telephony, television and data.

4.2 SYSTEM DEFINITION

It is necessary to consider two elements in the system equation, signal and noise, in order to calculate the performance of any earth station.

4.2.1 Noise input power

Equivalent noise input temperature and noise figure

The amount of noise in any communication system can be expressed in two ways; as an equivalent noise temperature measured relative to absolute zero with units in °K or as a noise figure which can be defined as the ratio of the input signal to noise and the output signal to noise, measured in dB. If we consider the basic noise equation:

$$W_n = \sqrt{4 \times k \times T_0 \times B_w \times R_l} \qquad (4.1)$$

where R_l = resistive load

        W_n = noise power across R_l

        B_w = bandwidth in Hz

        T_0 = actual ambient temperature

        k = Boltzmann's Constant

The noise parameters of an amplifier stage are defined in Fig. 4.2.

Any receiving system is comprised of a number of serial elements, shown in Fig. 4.3, each contributing some noise to the system (unless you can hold the element at a temperature of absolute zero) and thus degrading the required signal. In order to define equivalent noise temperature of a system each of the individual noise elements is referred to the system front end thus giving a total effective input noise of:

$$T_t = T_1 + \frac{T_2}{A_1} + \frac{T_3}{A_1 \times A_2} + \frac{T_4}{A_1 \times A_2 \times A_3} \qquad (4.2)$$

where T_1, T_2, T_3, T_4 etc. are the element noise temperatures and A_1, A_2, A_3

N_i = Input noise
A = Amplifier gain
T_2 = Amplifier noise temperature

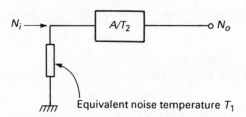

Fig. 4.2 Noise parameters of amplifier stage.

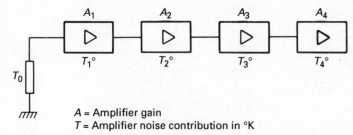

A = Amplifier gain
T = Amplifier noise contribution in °K

Fig. 4.3 Receive chain noise temperature.

etc. are the elements' respective gains. The relationship shown in equation 4.2 highlights the two important design factors that must be taken into account when designing a low noise receive system; first the input noise figure must be as low as practically possible as it is the major element in the system noise equation, second the first system element must have as high a gain as possible in order to minimise the effects of the noisy elements that follow the input stage.

In practical systems it is impossible to interconnect individual circuit elements with a zero loss element; this loss, depending on the type of interconnection such as cable or waveguide, increases as the distance between each element increases. Fig. 4.4 shows the loss element between antenna and receiver input and takes into account both external and internally generated noise.

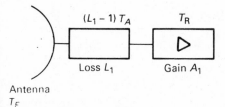

Fig. 4.4 Effect of loss element.

The equivalent noise temperature of the lossy element is given by equation 4.3:

$$T_1 = T_0 \left(1 - \frac{1}{L_1}\right) °\text{K} \qquad (4.3)$$

It should be noted that each system must be correctly matched at both input and output ports as any mismatch will appear as a lossy element in the total system noise.

4.2.2   Carrier power at the satellite

The signal power at the satellite receiver, P_{sr}, is defined by equation 4.4.

$$P_{sr} = (W_{txe} - L_{fu} - A_a - A_{sag} - L_{sf}) \qquad (4.4)$$

where W_{txe} = effective transmit power from earth station

   L_{fu} = free space loss

   A_a = atmospheric attenuation

   A_{sag} = satellite antenna gain

   L_{sf} = satellite antenna feed losses

4.2.3   Carrier power at earth station

The signal power at the earth station receiver, P_{er}, is defined by equation 4.5.

$$P_{er} = (W_{txs} - L_{fd} - A_a - A_{eag} - L_{ef}) \qquad (4.5)$$

where W_{txs} = effective transmit power from satellite

   L_{fd} = free space loss

   A_a = atmospheric attenuation

   A_{eag} = earth station antenna gain

   L_{ef} = earth station feed losses

4.2.4   Free space loss

The free space loss is calculated from equation 4.6.

$$L_p = 10 \log \left[\frac{4\pi d}{\lambda} \right]^2 \qquad (4.6)$$

where d = distance from satellite to earth station

λ = wavelength in cm

Fig. 4.5 shows the relationship between the free space loss and the distance for various frequencies.

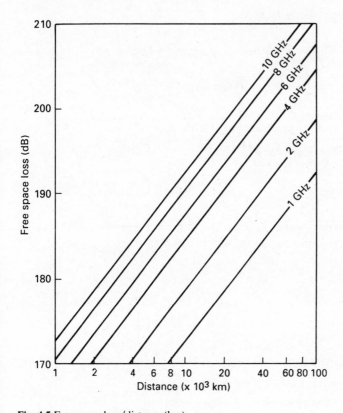

Fig. 4.5 Free space loss/distance (km).

The distance to the satellite depends upon the earth station position on the earth's surface, relative to the satellite. In chapter 2, Fig. 2.5 allows the designer to calculate the actual distance to slant angles up to 90°, though the minimum elevation angle is usually set to 5° by the regulatory authorities to minimise terrestrial interference and noise power injected into the receive antenna from the earth itself.

4.2.5 Satellite carrier to noise ratio

The satellite C/N is given by equation 4.7.

$$\frac{C}{N_s} = P_{sr} - 10 \log(k \times T_s \times B_{ws}) \qquad (4.7)$$

where T_s = satellite noise temperature
$\qquad\;\; B_{ws}$ = satellite Rx bandwidth

4.2.6 Earth station carrier to noise

The earth station carrier to noise is given by equation 4.8.

$$\frac{C}{N_e} = P_{er} - 10 \log(k \times T_e \times B_{ew}) \qquad (4.8)$$

where T_e = earth station noise temperature
$\qquad\;\; B_{we}$ = earth station Rx bandwidth

4.2.7 Downlink carrier to noise

The downlink carrier to noise is given by equation 4.9, which takes account of up and downlink noise and intermodulation distortion.

$$\frac{C}{N_d} = \frac{C}{N_e} + 10 \log(1 + r) \qquad (4.9)$$

where $r = T_{up} + \dfrac{T_{im}}{T_{do}}$

where T_{up} = up path noise temperature (°K)
$\qquad\;\; T_{im}$ = intermodulation noise temperature (°K)
$\qquad\;\; T_{do}$ = down path noise temperature (°K)

4.3 EARTH STATION COMPONENTS

4.3.1 Antennas

Introduction

The antenna requirements of a satellite earth station make it one of the most important elements in the overall earth station design; although non-electronic it contributes a major portion of the transmit and receive system gain, collects the microwave energy transmitted from the satellite, while discriminating terrestrial-borne interference. Its satellite equivalent can be used to concentrate the transmitted signal so as to illuminate any specific geographical area required. The antenna is required to have high gain in the forward direction and low gain in any other direction. This forward gain must be maintained regardless of elevation angle, wind velocity, temperature and extreme environmental conditions of heat, ice, and snow. The effective gain of any antenna system is not only a function of the reflector design but also of the pointing accuracy which the overall antenna structure is held to, a problem that is more acute for large antennas due to narrow forward beamwidth. In modern systems, where frequency re-use is in operation then the antenna has to discriminate between orthogonal polarisations. This discrimination is necessary to ensure minimum corruption of the received signal. The other major factor that has to be considered is the sidelobe performance, in which the sidelobes must be as small as possible in order to reduce interference with other satellites. 't is therefore of tremendous importance in obtaining the best orbital utilisation possible.

There have been a number of antenna systems used in satellite communications, all using some form of reflector and feed mechanism. Early systems used front fed parabolic antennas, which provided reasonable efficiency, with good sidelobe and gain performance. However they suffered from two major problems; high antenna noise (55 °K for an 85 ft (25.9 m) reflector) at 30° elevation and high feeder losses due to the waveguide losses incurred in taking the signals to and from the front feed. Another system sometimes employed is a horn antenna, which can give 70% efficiency and have a low antenna noise, but suffers from two major drawbacks; it is expensive and it has to be protected from the weather. In the early days two such antennas were built at Andover in Maine and Plemair Bodou in Brittany and although the design was further developed by folding the horn in order to reduce the physical length, this reduction was only accomplished at the expense of antenna efficiency and the design made no further progress.

The antenna configuration most commonly used throughout the world is the Cassegrain system, whose general form is shown in Fig. 4.6. This uses a main parabolic reflector with a hyperboloid sub-reflector positioned at the focal point of the parabola. This system reduces the loss associated with the front fed systems as the feed is situated at the vertex of the paraboloid.

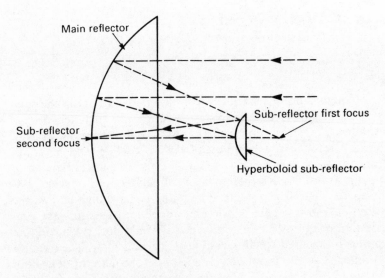

Fig. 4.6 Basic Cassegrain antenna geometry.

Antenna gain

The antenna gain is a function of the effective aperture area, which takes account of the antenna losses due to the inaccuracies in the main and sub-reflector profiles. The gain is expressed relative to that of an isotropic antenna and is given by equation 4.10.

$$G = 10 \log \text{ dBi} \qquad (4.10)$$

For a lossless antenna the effective aperture area is given in equation 4.11.

$$A = \frac{(\pi \times D)^2}{4} \qquad (4.11)$$

where A = aperture area
D = diameter of paraboloid

Thus the effective antenna area is:

$$A_a = nA \qquad (4.12)$$

where A_a = effective aperture area
n = antenna efficiency

To calculate the maximum gain G_m:

$$G_m = \frac{4\pi \times A_a}{\lambda^2} \qquad (4.13)$$

where λ is wavelength in metres.

Expressed in decibels the gain becomes as in equation 4.14.

$$G = 9.94 + 10 \log_n + 20 \log \left(\frac{D}{\lambda}\right) \text{ dBi} \qquad (4.14)$$

It is important to note that the antenna efficiency is not only a function of theoretical antenna efficiency and the feed losses. The relationships quoted above apply not only to the transmit but also the receive mode of operation. The reflector accuracy is also a factor affecting the antenna gain and at the same time increases the sidelobe power. The loss of gain due to inaccuracies is given by:

$$dG = 0.00761(e \times f)^2 \text{ dB} \qquad (4.15)$$

where e = surface error in mm

f = frequency of operation in GHz

This gives 0.92 dB loss at 11 GHz for 1 mm of surface error.

The surface accuracy of the antenna reflector is made up of a number of static and dynamic factors. The reflector surface manufacturing accuracy is determined by the tool or mould tolerance and where the parabola is multi-sectioned then the setting up accuracy must also be taken into account. The dynamic distortions are due to the external elements such as wind and snow plus the fact that the weight of the antenna itself causes the antenna to distort simply due to gravity effects.

Antenna pointing

Up till now the system gain requirements have simply been treating the antennas as a theoretical sub-system but it is also necessary to take into account the effect of the practical antenna characteristics on the effective system gain.

The antenna must maintain its performance, regardless of the environmental conditions in which it has to operate, in terms of its pointing accuracy relative to the satellite and the tracking of any satellite movement. In the early days of satellite communications the satellite maintained its position, relative to the earth station to an accuracy of 0.1° though this is now reduced due to improvement in the satellite station-keeping capability. The pointing accuracy of the antenna has to be held in order to ensure that the power transmitted to and received from the satellite is kept to within the system parameters.

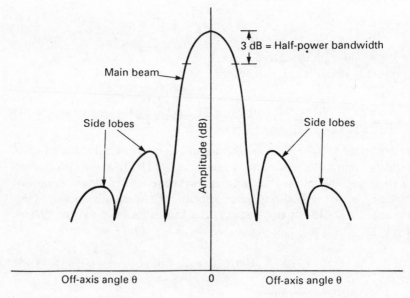

Fig. 4.7 Paraboloid antenna radiation pattern.

The pointing accuracy requirements are dependent upon the radiation pattern shown in Fig. 4.7.

Obviously if the main beam bandwidth is wide enough, i.e. a small reflector diameter, then there is often no need to have a tracking system to follow the satellite movement and therefore a simpler antenna structure can be used. The antenna beamwidth of θ_{3dB} is inversely proportional to the operating frequency and antenna diameter and is derived from equation 4.16 which shows the directivity of the antenna.

$$\theta_{3dB} = \frac{k\lambda}{D} \tag{4.16}$$

where λ = wavelength in mm

K = aperture dependent constant

D = antenna diameter

The value of K for high efficiency antennas is 70. This means that for a 32 m diameter antenna, using a 6 GHz transmitting frequency, the 3 dB (half power bandwidth) is 0.11°.

There are two other factors that must be taken into account when considering the need for tracking and these are the orbital inclination of the satellite and the location of the earth station; orbital inclination increases the satellite drift and although the satellite inclination is usually less than 0.1°, antenna designs often take 1.0° as a design parameter. A typical design, assuming a 1.0° orbital inclination, would give an elevation velocity of 0.004°/minute and an azimuth velocity of 0.0049°/minute.

The step track system has taken a long time to mature as in the early days it was considered to be too inaccurate for professional use. The principle of step track is simple, the antenna is moved in a set pattern relative to its original position and the change in amplitude measured. If the amplitude increases then this process is continued until the maximum is reached when the search process is stopped.

The problem with this method is that the system cannot easily distinguish between signal step and noise and in its simpler form it cannot handle rapid changes of satellite and/or antenna position due to wind perturbation. However the great advantage of this method is its simplicity and cost effectiveness; because of this it has been refined, by the use of sophisticated digital processing, to produce what is known as a smoothed step track. Here the receiver builds up an overall picture of the satellite's movement and stores it, then in the autotrack mode the microprocessor generates scan commands that displace the antenna in a set pattern from the memorised satellite position. This routine determines the size and direction of any error and once on boresight it is held in position. The maximising process is not performed continuously but is only activated when the received signal falls below a set threshold or after a fixed cycle time. The use of an intermittent interrogation cycle allows for operation in steady state conditions of wind and satellite, without antenna perturbations. The cycle time is automatically adjusted to take account of any changes in weather conditions or satellite movement. The pattern used is shown in Fig. 4.8. The origin represents the last position of the satellite, the processor commands the antenna to move between A, B, C and D in turn; the azimuth information is the integration of

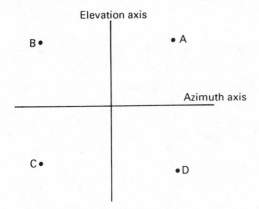

Fig. 4.8 Step track movement pattern.

the inverse of the signal levels at points B and C. If the satellite is correct in its azimuth position the integral of the signals at A and D cancels the inverse of the signal derived from the integration of B and C to give a zero output to the tracking receiver; for any elevation error the difference is not zero and signal output is proportional to the pointing error, the elevation error is derived by comparing the integrated total of A and B and the inverse of D and C.

The control signals, derived by any method of tracking, are used to drive a closed loop control system which takes axis position data derived from position transducers coupled to each axis. The antenna is moved in azimuth or elevation by the use of DC motors. The tracking system not only has to remove pointing error but must also counteract the mechanical characteristics of the antenna, such as inertia, gearing backlash and friction.

Sidelobe performance

The antenna radiation pattern shown in Fig. 4.7 has to be defined in precise terms to ensure that the sidelobe radiation does not interfere with the satellite or any terrestrial communication systems.

The pressure on the use of the geostationary orbit has led to the tightening up on the regulations associated with the sidelobe parameters, in order to allow a 2° spacing between satellites. The CCIR Recommendation 465 specified a reference radiation diagram as follows:

$$G = 32 - 25 \log \theta \text{ dBi} \quad \text{where} \quad 1° < \theta < 48°$$
$$G = -10 \text{ dBi} \qquad \text{where} \qquad \theta > 48° \tag{4.17}$$

where G is the gain in dB, relative to an isotropic antenna, in the direction of the satellite and within 1° north and south of the orbit. The antenna characteristics, defined for early antennas, allowed a 3° spacing; in order to have satellites at 2° spacing the satellite antenna radiation pattern is now defined as:

$$G = 29 - 25 \log \theta \text{ dBi} \quad \text{where} \quad \theta < 20° \quad \text{and} \quad \frac{D}{\lambda} > 100 \tag{4.18}$$

With the above the sidelobe performance must be met by 90% of all of the sidelobe peaks beyond an off-axis angle of 1°. These patterns are similar to that specified by the Federal Communication Commission (FCC) which defines more precisely the radiation pattern between 1° and 48° as below:

$$\begin{aligned} G &= 29 - 25 \log \theta \text{ dBi} \quad \text{where} \quad & 1° < \theta < 7° \\ G &= 8 & 7° < \theta < 9.2° \\ G &= 32 - 25 \log \theta & 9.2° < \theta < 48° \\ G &= -10 & \theta > 48° \end{aligned} \tag{4.19}$$

These figures will be superseded in the future when the radiation patterns will be rigorously enforced for all sidelobes and the off-axis angle decreased to 36°. In current designs, using off-set feed configurations, the regulatory specifications are now being exceeded by up to 6 dB, allowing either a superior performance or decreased transmitter power to be achieved.

The increasing importance of antenna performance means that the design must ensure that the reflector and sub-reflector profiles are engineered to very fine tolerances and certain antenna characteristics tailored for maximum performance.

Spillover effects

If we consider a standard Cassegrain antenna system then the transmitted signal is directed towards a sub-reflector, situated at the focal point of the reflector and ideally all of that forward energy, which is in the direction of the satellite, would be reflected from the sub-reflector back to the main reflector before being radiated to the satellite.

Antenna tracking systems

To ensure that the antenna is held on boresight, i.e. the satellite, a variety of systems have been devised to achieve this function. The method will have to ensure that the antenna can follow the satellite at its maximum velocity, while ensuring that its own errors, due to wind, servo and motor drive backlash are minimised and remain within the error margin of the system.

There are four main forms of tracking system:

Conical scan
Conical scan was used in the early 1970s as a tracking method but being an active mechanical method of tracking it has a number of problems. The basis of its operation was that it rotated the sub-reflector at a low frequency which had the effect of rotating the main antenna beam around the normal boresight; this rotation, at a low frequency of around 400 Hz, modulated the tracking beacon signal as it was transmitted by the satellite, with modulation varying in amplitude and phase depending upon the antenna pointing error. This modulation can be recovered to provide error control signals that are then used ꞏo drive the antenna servo system. The problems oᶠ ꞇhis method are first that the modulation was not only on the beacon but also on the transmit and receive signals and therefore had to be kept to a low value with consequent problems in demodulation and second the modulating frequency had to be set well above 50 Hz to avoid hum problems but the higher the frequency the more the loss of effective antenna gain.

Monopulse systems
This method is able to derive tracking signals from a single received signal by

use of a specially derived feed. One method has a four-horn system arranged in a square, with feed focus at the centre of the square, and the output of the four feed horns combined to give a sum signal and signals proportional to the horizontal and vertical displacement of the antenna from boresight. The second method utilises the fact that higher order modes will be generated in the feed waveguide when the antenna is off boresight and these modes give phase and amplitude information that can be used to produce error signals that will reposition the antenna. The four-horn method is superior to the mode system in that it maintains the independence of the received polarisations, a very important characteristic when the earth station is operating in a re-use mode, while the waveguide mode is sensitive to signal depolarisation and can therefore produce antenna pointing errors.

The other drawback in using a monopulse system is the cost involved in designing complicated feed systems and complex tracking receivers and it is therefore used when a very high performance system is required i.e. a 32 m antenna system where the high cost is a lower proportion of the overall earth station cost.

Step track system

The alternative to conical scan or monopulse is a step track that is analogous to a conical scan system without a rotating feed or sub-reflector. If the signal is not completely reflected then the non-reflected signal spills over, as an interfering signal into the main beam. This sub-reflector spillover not only affects the antenna radiation pattern but also appears as a noise contribution at the input to the earth station receiver, which for large Cassegrain structures is in the order of 7 °K. The same considerations apply to main reflector spillover, which produces radiation fields behind the main reflector, thus producing problems of interference with terrestrial systems. The effect of spillover in terms of sidelobe performance depends upon the angle off boresight and is not a major factor for off-axis angles of less than 5°, but becomes significant beyond this angle and is a dominant factor beyond 90°.

There are a number of methods for reducing the effects of spillover:

(a) The reflector and sub-reflector can be designed with non-uniform profiles to reduce edge illumination.
(b) The sub-reflector can be fitted with a guard between feed horn and reflector, an alternative that could only be countenanced in an offset antenna configuration due to the sub-reflector blockage by this method.

Signal blockage

The physical construction of the antenna itself produces barriers to the microwave radiation. The sub-reflector, or feed, if front fed, together with its supports interfere with the transmission path between the reflector and the satellite, thus producing what is known as aperture blockage. In addition to the design of the sub-reflector the position of its supports is very important as they can interfere with the illumination of the sub-reflector or feed. These effects result in a reduction in the antenna gain and an increase in the sidelobe amplitude relative to the main beam. The degradations can be reduced by optimum sizing of the sub-reflector and optimal placing of sub-reflector supports or the use of an offset system.

Signal scattering

Any metallic element, within the main beam of the antenna actually becomes a radiator in its own right and therefore adds in to the sidelobe energy, as well as increasing the noise temperature of the antenna. This scattering can be reduced by the minimising of the physical elements, reducing the actual excitation of the physical element by the use of microwave absorbent material placed around the struts etc. and the attaching of absorbent material to the outer rim of the main reflector.

Effect of sub-reflector tolerances on sidelobes

The deviation from the normal design parameters of the reflectors produces a variety of effects, not just on the sidelobe performance but also on the gain of the antenna. The mechanism by which the surface errors produce reduced performance is that surface errors generate phase errors across the antenna aperture, which increases the sidelobe energy and reduces the effective gain of the antenna.

The effect on antenna performance was set out by Ruge in 1966 and his theory allows the sidelobe power to be predicted against the actual profile, as well as calculating its effects on antenna gain. The actual performance depends upon the size of correlation interval provided that the random errors are above 0.2 mm rms. In practice it is very important to ensure that the reflector is smooth over individual areas rather than maintaining the overall accuracy across the total surface.

Cross-polarisation performance

The performance of antennas when frequency re-use involves the transmission of horizontal and vertical polaristions, must obviously be considerably better than for a single case. The Cassegrain system provides good cross-polarisation discrimination on axis but deteriorates once the antenna's pointing angle to the satellite is no longer in the plane of the geostationary arc. It is therefore necessary to limit the cross polarisation both on and off axis. The effect can be reduced by the use of a paraboloidal sub-reflector which cancels out unwanted signals.

Offset fed antennas

The question of antenna size reduction has always been a major consideration in earth station design, both on cost and environmental grounds. In the beginning the major limiting factor was the satellite itself, which was weight-limited and therefore restricted in power output and overall life span. The first INTELSAT earth stations, for trunk communications purposes, had to use 32-metre reflectors in order to achieve the required figure of merit, G/T, a limitation not only due to the satellite characteristics but also because the low noise amplifiers, LNA, were limited in performance and needed to use cryogenics in order to achieve even that performance. A typical LNA had a gain of 50 dB and an effective input noise temperature of 60 °K at 4 GHz. The trend now is towards smaller antennas, brought about by the use of higher frequency bands, improved satellite technology that allows greater signals to be transmitted from the satellite and the availability of smaller, more cost effective LNAs that can provide noise figures of 180° at 11 Ghz. Indirectly this has led to the development of small, efficient antennas having approximately 44 dB forward gain.

The dual reflector antenna system, whether Cassegrain or Gregorian, can give a much improved performance if the sub-reflector is taken out of the main reflector aperture; the improved performance encompasses antenna gain, sidelobe characteristics and cross-polarisation levels. A typical system is shown in Fig. 4.9. This configuration uses an ellipsoidal sub-reflector

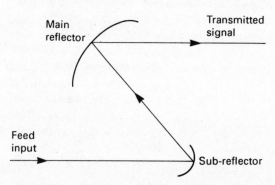

Fig. 4.9 Offset Gregorian antenna configuration.

which allows a feed position that is on the axis of symmetry of the paraboloid and close to the edge of the main reflector. In addition the horn aperture only needs to be a few wavelengths long. The main reflector can be either a square, circular or elliptical parabaloid. It has been found that the elliptical reflector can give an improved rate of fall in sidelobes, in both the major and minor axes for the same gain, which is very much a function of surface area.

The antenna must be physically configured to give optimum cross-polar performance in order to counteract the cross-polar signals generated by the offset arrangement, leaving two further elements of cross-polar distortion, diffraction and paint layers on the reflector surfaces.

One major advance in reflector design has been the development of diffraction profile synthesis, DPS, which can give very high aperture efficiency and resolves the conflicting requirements of high aperture illumination and low main reflector spillover. An alternative approach, using square main reflectors, is to design the feed to include a dual polarised mode generator which actually cancels the cross-polarisation effects to give very low sidelobes, well below the regulatory limits.

Antenna feed system

The feed system should ideally couple directly into the low noise system of the earth station receiver and, in addition, must be capable of handling the required transmit power with as little loss as possible.

4.3.2 Low noise amplifiers

The low noise amplifier system is a crucial element in the receiver chain of any satellite earth station, circuit failure would render the system unworkable. One of the methods for ensuring that the system has a high availability is to build reliability into the components of the amplifier, though this is not a sufficient safeguard when the loss of traffic means a loss of revenue. The major method is to use redundancy techniques to safeguard the most important elements of the earth station. The standard method is to use a 1 for 1 configuration, with a main and standby unit connected as shown in Fig. 4.10. The system is arranged so that one of the LNAs is designated as the main amplifier, through which the received signal is amplified, with the

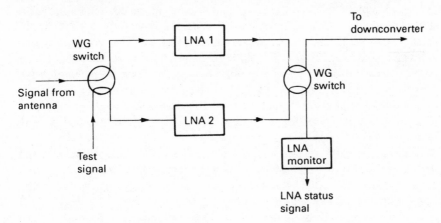

Fig. 4.10 1 for 1 redundancy LNA system.

second amplifier being operated in a hot standby mode. The hot standby is fed with a test signal that is used to check its functionality so that should the main system fail the supervisory system can initiate a changeover to the second amplifier and so prevent overall system failure. The mon˙ ɹring of the amplifier can be relatively unsophisticated, simply monitoring the LNA supply status, or it can use a test translator system, as shown in Fig. 4.11.

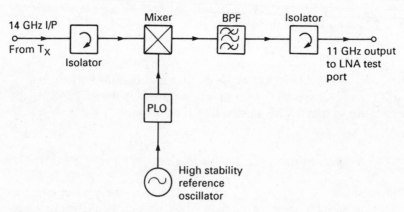

Fig. 4.11 14 GHz/11 GNz test translator.

The test translator input signal is derived from the earth station high power amplifier output, which is translated into the receive frequency band. Where more than one polarisation is received and the cost of the earth station is of paramount importance the form of redundancy is reduced to 1 for N, which reduces the amount of equipment needed and therefore the cost.

LNA amplifiers have developed rapidly over the past few years, mainly due to commercial pressures for smaller and more cost-effective earth stations, both for business communications, TV receive only (TVRO) applications and very small aperture terminals (VSAT). The development of gallium arsenide (GaAs) FETs as a microwave element has meant that amplifiers can be devised that will operate up to 40 GHz. These monolithic microwave integrated circuits (MMIC) can provide compact units that not only provide input noise temperatures of 100 °K but also a downconversion to a lower frequency of say 1 GHz, which is more easily handled by any receiver.

Any satellite earth station has to meet its specified figure of merit for the service in which it is going to be used. This figure of merit involves the system noise temperature, whose major component is the noise temperature of the LNA, which should have as low a noise figure as possible and as high a gain as possible in order to minimise the effects of the following stages.

As stated earlier the first earth stations utilised cryogenic systems to obtain a low noise temperature. They were however very large in size and weight and had to be mounted in a special enclosure directly behind the antenna feed in order to minimise the losses and prevent the increase in noise temperature due to those losses. The enclosure also performed another function which was to provide a constant temperature for the LNA. The problems of such a system were resolved initially by designing an almost lossless means of connecting the feed output to the LNA, i.e. a beam waveguide system that transferred the signal by means of a series of mirrors from the feed output to the LNA input, which could now be mounted in the earth station some distance from the antenna.

Modern LNAs, for the larger earth stations, utilise thermo-electric cooled parametric amplifiers which can give a noise figure of 45 °K and a gain of 50 dB at C Band, though at Ku Band the noise figure attainable is in the order of 150 °K. The actual effective noise temperature is, as already stated, in part determined by the earth station layout. In certain instances the antenna building is separate from the ground communication equipment room in which case the signal can either be conveyed across site at micro-wave frequencies or translated down to a standard IF frequency at 70 or 140 MHz.

12 m Parabolic
antenna

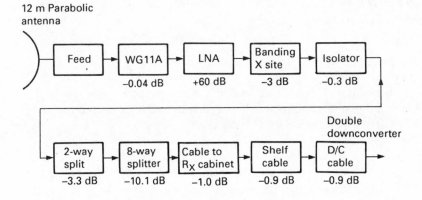

Fig. 4.12 Loss/gain elements in INTELSAT B earth station.

A typical system, involving an INTELSAT Standard B earth station is shown in Fig. 4.12. This system shows all the elements that a typical layout would have. The output from the 12 m antenna feed is connected to the LNA by the section of waveguide WG11A which has a very low attenuation. The LNA has a total noise figure of 75 °K, which is not simply for the amplifier but takes into account the fact that the LNA is normally configured in a 1 for 2 redundancy mode, that utilises a special mounting plate assembly. The LNA is then connected via cable, whose length is determined not only by the cross-site distance but also by the banding cable which is needed to take account of the antenna movement in azimuth. The signal is isolated, which introduces a loss, before being split out to a number of receiver chains for telephony, data and television.

In order to calculate the signal input to the double downconverter system it is essential to take into account the inter-cabinet and shelf cabling as shown in Fig. 4.12.

Standard B figure of merit

The design parameter for the Standard B earth station is as equation 4.20.

$$\frac{G}{T} = 31.7 + 20 \log \frac{f}{4} \text{ dB/K}° \qquad (4.20)$$

The figure of merit above is the minimum requirement, which must be met with the antenna pointing at the satellite, under clear sky conditions and for any frequency in the 3.7–4.2 GHz band. The effective antenna gain at 4 GHz is derived from:

Nominal gain of 12 m antenna	53.1 dBi
Guaranteed gain	52.9 dBi
Feeder waveguide loss	− 0.4 dB
Mismatch loss	− 0.15 dB
Nett gain	52.7 dBi

The figures quoted also bring out one other parameter that must be minimised and that is the mismatch between units and their interconnection. If this is ignored then the G/T will be reduced.

System noise
This must take into account antenna noise, LNA input noise and plate assembly contributions, together with second stage noise referred to the input and also mismatch effects. Thus we have a total noise of:

Feeder waveguide loss and mismatch	9.7 °K
Antenna/LNA mismatch	2.0 °K
Second stage contribution	1.0 °K
LNA/plate assembly	75 °K
Antenna noise/10° elevation	32 °K
Total	119.7 °K

The effect of elevation on antenna noise is shown below:

Antenna noise temperature/elevation

Elevation	Noise °K
10°	32
20°	23
30°	22
40°	20

Figure of merit calculations

C Band earth station G/T

$$G/T = \text{Gain dBi} - 10 \log (119.7)$$
$$= (52.7 - 20.8) \text{ dB/°K}$$
$$= 31.9 \text{ dB/°K}$$

This is the minimum figure on the lowest antenna gain and elevation. If we take the operating elevation, in this case the Atlantic Ocean Satellite from Mauritius, then there is a 0.8 dB margin above the statutory minimum. The other assumption made in the above calculations is that the LNA is mounted in the antenna hub and held at a reasonably constant temperature; if this is not the case then as the ambient temperature increases the figure of merit will decrease.

Ku Band earth station G/T

It is useful to consider the figure of merit for a Ku Band system in that it shows the order of magnitude of the various elements in the receive chain (see Fig. 4.13).

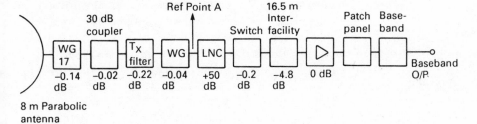

Fig. 4.13 Ku Band receive chain loss and gain elements.

The waveguide used at 11 GHz is WG 17, having a 0.159 dB/m loss, which should be kept to a minimum to optimise the figure of merit. The noise contributions of these stages, following Ref Point A, have to be referred back to that point.

Ku Band system performance

Gain

Antenna gain	57.5 dBi at 11.2 GHz
Waveguide loss	0.38 dB
Mismatch loss	0.14 dB
Effective gain	56.98 dBi

Noise (25° elevation)

	Parameter	°K at A
Antenna noise	63 °K	57.72
Waveguide	0.38 dB	24.3
Mismatch	6.1 °K	6.1
LNC (23 °C)	2.3 dB (NF)	202.5
Second stage noise	21.3 dB (NF)	0.39
Total noise		291.01

This gives a figure of merit of 32.34 dB/°K. At 50 °C the G/T will reduce to 32.05 dB/°K due to degradation of LNC performance.

4.3.3 High power amplifier (HPA) systems

Introduction

The power amplifiers used in satellite earth stations are a critical element in the communication system. They directly affect the quality and cost of any transmission, which means that there are certain parameters that have to be optimised in the amplifier design. The amplifier must not add distortion to the signal and therefore must have a linear transfer characteristic, minimal gain slope (dB/MHz) and low AM/PM conversion. If an amplifier is operated near to saturation and is transmitting multiple carriers then any non-linearity causes AM/AM conversion and AM/PM conversion between the multiple carriers. The effect of this is twofold; there is intermodulation between the two carriers as shown in Fig. 4.14.

The intermodulation frequencies are determined by the two carrier frequencies and the level of distortion is a function of amplifying tube non-linearity and the spread of the spectrum of each carrier. The effect of intermodulation is to effectively increase the system thermal noise but it does not effect the noise in the baseband.

The presence of non-linearity in the amplifier will produce distortion in the baseband due to intelligible crosstalk, which is produced when two FM carriers are transmitted through an amplifier that has a non-linear gain/

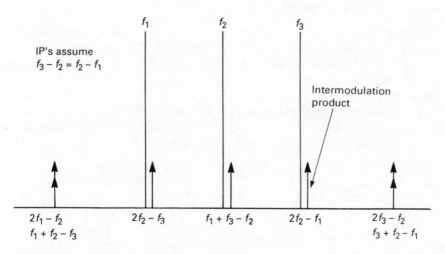

Fig. 4.14 Third order intermodulation products.

slope characteristic. The effect can obviously be minimised by ensuring that the amplifier has a low AM/PM conversion factor and minimal gain/slope.

Intelligible crosstalk will manifest itself in FM systems regardless of whether it is a telephony or a television service. For FM multiple carrier systems the intelligible crosstalk can be calculated from:

$$I_{xt} = 20 \log \left[\frac{180}{\pi} \times \frac{1}{K_p \times G_s \times f} \times \frac{f_r}{f_{ri}} \times \frac{P_t}{P_i} \right] \qquad (4.21)$$

where I_{xt} = TT Level/Xtalk ratio

K_p = AM/PM conversion coefficient °/dB

G_s = system gain/slope dB/MHz

f = frequency of crosstalk

f_r = test tone deviation of carrier Hz

f_{ri} = rms frequency deviation of interference fMHz

P_t/P_i = total carrier power/interfering carrier power

For television systems it is probable that two video signals would be transmitted and non-linearity would produce chrominance crosstalk, with baseband chrominance information being transferred between carriers. This effect is frequency dependent and this manifests itself more at the top end of the baseband.

Types of high power amplifier

There are three basic types of high power amplifiers in common use in satellite earth stations, each type being characterised by the type of output device in use.

Klystron tube amplifiers
The klystron amplifier has been the major form of HPA since the inception of satellite communications, particularly for multi-carrier and television carrier transmissions. The basic form of the amplifier is shown in Fig. 4.15.

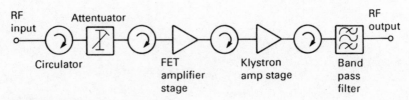

Fig. 4.15 Klystron HPA configuration.

The main elements in the amplifier are the klystron output tube and the low level driver, which is usually an FET amplifier. The klystron tube can provide cost-effective power amplification for power outputs between 750 W and 10 kW. The common frequencies of operating are from 4–18 GHz and an amplifier bandwidth is typically 0.75% of the transmission frequency and the centre frequency can be tuned across approximately 10% of its value. The FET amplifier must be capable of driving the main tube into saturation without introducing intermodulation distortion into the signal, i.e. it must have a high third order intercept point. In addition it must have a reasonable noise figure in order that minimal noise is introduced into the HPA transmission band.

The klystron amplifier has been used in C Band trunk systems, providing up to 3 kW of output power over a frequency band of 5850–6425 MHz. The other elements in the HPA are required in order that the system will be safe and efficient in operation. The facilities required are:

- Isolation of the amplifier elements to protect against reflected power;
- Control of output power;
- Muting of the amplifier output should fault conditions, such as power arcing, occur;
- Forward and reverse monitoring by use of couplers;
- Harmonic filtering at the output.

The amplifier construction must be as compact as possible in order to minimise the length of the interconnections between amplifier elements. This means that isolators, waveguide and filters must be designed for minimum insertion loss because any loss will actually increase the intermodulation level. The major parameters that have to be considered in any HPA design are:

AM/PM conversion

For klystrons the AM/PM conversion is typically 4°/dB at saturation, reducing to 2.5°/dB when the amplifier output is backed off by 3 dB. This would give an overall intermodulation product level (IP) of − 10 dB/ − 20 dB respectively. The IP level at other output powers for two equal carriers is shown in Table 4.1.

**Table 4.1   Effect of power O/P on
intermodulation distortion**

Power O/P	IP product/dB
Saturation	− 10
− 3 dB	− 20
− 10 dB	− 34
− 20 dB	− 54
− 30 dB	− 74

The importance of the output level being non-saturated is clearly shown. It should also be noted that the major element in the system intermodulation is due to the klystron tube itself.

Gain/slope dB/MHz

As already discussed this parameter is critical in the intermodulation distortion and intelligible crosstalk performance of the amplifier.

Group delay ns/MHz

This parameter is usually restricted to a defined performance and is given in the form of a mask that sets the parameter over the operational bandwidth.

Noise

This parameter is not only a function of the tube performance but also the characteristics of the other amplifier elements.

Third order intermodulation products

Harmonics/spurious

Both of these elements must be kept to a minimum, typically − 80 dB relative to the carrier.

Travelling wave tube amplifiers

The klystron amplifier is only capable of transmitting a limited number of carriers and should the frequency of operation need to change then the klystron amplifier usually has to be returned, an operation that can be carried out automatically but which takes a finite amount of time and therefore makes TDMA operation more difficult with klystron amplifiers.

The problems can be overcome by the use of a travelling wave tube system that has a much wider bandwidth. The total amplifier configuration is the same as that shown in Fig. 4.15 except that the main output element is replaced by a travelling wave tube. It can give a bandwidth of up to 10% of the transmission frequency, i.e. 500 MHz at C Band. The problems of intermodulation distortion still remain but can be significantly reduced by the use of a lineariser which is inserted at the input to the amplifier and provides a gain-slope characteristic that is equal and opposite to the TWT performance. This linear characteristic can reduce the IP products by 10 dB. The TWT amplifier is unable to provide the range and magnitude of the klystron tube, it is however in wide use in modern earth stations, which require less power output, and provide outputs up to 800 W with the additional advantage of smaller physical size and weight.

Solid state amplifiers (SSAs)

The use of solid state components, to provide earth station or satellite power amplifiers, appears to be a very attractive alternative to the use of travelling wave tubes (TWT) systems. The possible advantages are:

- Smaller size;
- Higher reliability;
- Smaller power supply requirements;
- Lower costs.

These are not yet realities, except at lower power output levels, but the SSA has problems of absolute power output level, amplifier gain-temperature instability and poor resistance to power supply transients.

The development of SSAs has to concentrate not only on increasing the amount of power output available but also the amplifier efficiency (currently below 20%), which will increase cost and power supply complexity.

The lower frequency systems, at L Band, can utilise bi-polar transistors, connected in amplifying cells as basic units, which are themselves paralleled together to provide up to 40 watts output at 1.6 GHz. At higher frequencies the MESFET transistor is becoming the major element in C, Ku and Ka Band power amplifier design. In combination with monolithic microwave integrated circuit (MMIC) techniques the use of MESFET transistors can currently provide 10 W output at C Band, with 23% efficiency. Over a bandwidth of 500 MHz, as the frequency of transmission increases, the power output performance reduces where amplifiers operating in Ku Band are likely to have only 3 W output. Apart from these problems, which should diminish as technology advances, the SSA is more linear than TWTAs, more reliable in current technology and is thus able to be used both in satellites and small VSAT systems.

HPA system redundancy

The failure of the HPA would mean that the earth station would become a one-way, receive-only system and in the majority of installations the high power amplifiers are arranged in a redundancy configuration. There are, however, a number of uses, such as low weight transportable units, which operate with only one power amplifier (on weight and cost grounds), where it is accepted that transmit system failure may take place.

The power amplifier system can be one of the most expensive elements in an earth station, especially trunk route systems which use large numbers of individual amplifiers. The redundancy system can be extremely complex, especially in multiple amplifier earth stations, where a 1 for 1 system would be inappropriate and too expensive. The simplest form, a 1 for 1 system is shown in Fig. 4.16.

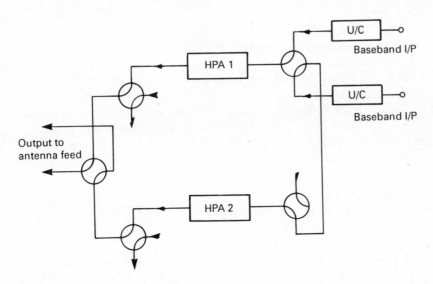

Fig. 4.16 HPA redundancy system.

In addition to the main amplifiers the drive system is often included in the switching arrangement. This method ensures that the complete transmit chain is correctly equalised for both amplitude and group delay.

HPA system reliability

The dominant factor in assessing any system is not just the individual items but the maintenance system within which it is operating and the spares holding. Table 4.2 shows the mean time between failure (MTBF) for transmit system components. The major difficulty in calculating MTBF is the basis on which component reliability is determined, in this instance a military standard, MIL–HDBK–217D, is used.

Table 4.2   Equipment reliability characteristics

Unit	MTBF (hours)	MTTR (hours)
HPA	6 000	1.0
U/C	18 000	2.0
WG Sw	50 000	2.0

The HPA is a 3 kW klystron, connected by waveguide switches and operated by some form of logic, whose MTBF must also be taken into account. It can be seen from Table 4.2 that the most unreliable single element is the amplifier itself.

4.3.4   Ground communication equipment (GCE)

The GCE equipment is used, either to process the baseband signal and upconvert it to the transmit frequency or downconvert the output of the low noise amplifier for further processing in order to recover the transmitted information. The common name for this type of equipment is ground communication equipment, usually housed away from the low noise amplifiers and the high power amplifiers.

Receive system
The basic elements of a receiver are shown in Fig. 4.17.

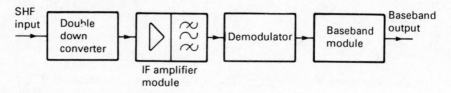

Fig. 4.17 Ground communication receiver chain.

Downconversion

The downconverters block diagram is shown in Fig. 4.18 and its function is to take any signal, within the satellite transmit frequency band, and to translate that input down to an intermediate frequency that is simpler to process.

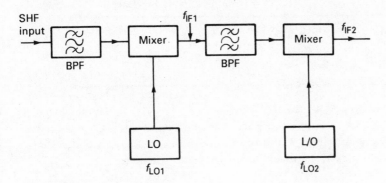

Fig. 4.18 Downconverter block diagram.

The input frequency band can be very wide, e.g. 600 MHz in C Band. This means that the downconverter must be capable of being set to the required input frequency band, either by a physical change of the first local oscillator or by the use of a frequency synthesiser as the first local oscillator, which can be set in fixed frequency steps.

It has to be recognised that the design of the downconverter is a complex process which generates unwanted spurious frequencies that fall within the required frequency band. The input band is set by satellite system allocations and the intermediate frequency bandwidth, centred around the final IF frequency, is normally 70 MHz or 140 MHz.

The local oscillator, a necessity for the downconversion process, also contributes noise and FM components to the receiver chain. In addition the stability of the oscillator is important in that it sets constraints upon the frequency stability of the overall receive chain.

The input filter, shown in Fig. 4.18 has a bandwidth that is suited to the system bandwidth required. The local oscillator frequencies, whether the first or second stage conversion, must be set such that:

$$F_{lo} = F_{rf} - F_{if} \qquad (4.22)$$

This local oscillator frequency, set below the input frequency, ensures that spectrum inversion does not take place. The mixer produces spurious components, derived from harmonics of the local oscillator and IF frequencies, as well as cross modulation products. The spurious outputs must fall within the frequency band given by equation 4.23.

$$F_{if} \pm \frac{B_{wif}}{2} \qquad (4.23)$$

In a typical system the IF frequencies would be as shown as in Table 4.3.

Table 4.3   Rx IF frequencies of typical systems

I/P Frequency GHz	1st IF MHz	2nd IF MHz
3.6–4.2	770 (2.8–4.3)	70 (700)
12.5–12.75	1200 (11.3–11.55)	140 (1.06)

The synthesiser local oscillator adjustment is in 125 KHz steps for a typical INTELSAT system. For a fixed local oscillator, adjustment is made by replacement of a high stability low frequency source, around 5 MHz, which is used to control a much higher frequency oscillator.

Transmit upconverters

The basic form of an upconverter is shown in Fig. 4.19.

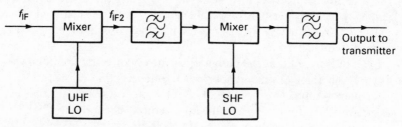

Fig. 4.19 Basic upconverter configuration.

In a typical system the system frequencies would be as shown in Table 4.4.

Table 4.4   Tx IF frequencies of typical systems

Tx frequency GHz	1st IF	2nd IF
5.85–6.42	770	70
14.0–14.5	1200 (1280–1330)	70 (1060)

System equalisation

The ground communication equipment usually contains IF modules that provide signal processing that ensures the correct output level to the demodulator system. However, more importantly the module has to correct the amplitude and group delay distortions caused by the receive and transmit path characteristics. A typical requirement is shown in Fig. 4.20.

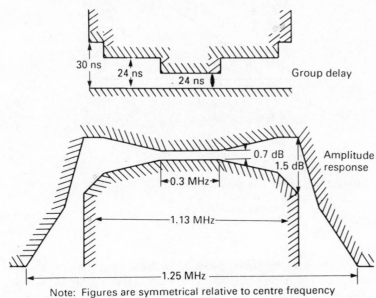

Note: Figures are symmetrical relative to centre frequency

Fig. 4.20 1.25 MHz carrier amplitude and group delay requirements.

Pre-emphasis/de-emphasis

In order to reduce the thermal noise and so improve the signal/noise ratio of the system the baseband signal to be transmitted is processed to modify the signal level distribution over the full range of baseband frequencies. The processing characteristics are shown in Fig. 4.21. These characteristics are derived from equation 4.24.

Relative frequency deviation of test tone at baseband frequency f

$$= 5 - 10 \log \left[1 + \frac{6.90}{1 + \frac{5.25}{\left(\frac{f_r}{f} - \frac{f}{f_r} \right)^2}} \right] \quad (4.24)$$

where $f = 1.25 \, f_{max}$

f_{max} = highest baseband frequency

This characteristic has to be maintained to within $\pm (0.1 + 0.5 f/f_m)$ dB. The effect of the pre-emphasis characteristic is to increase the frequency

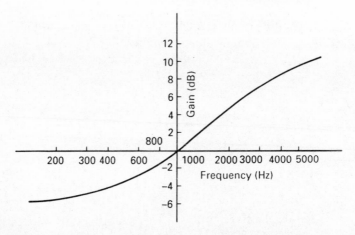

Fig. 4.21 Pre-emphasis characteristics.

deviation of the signal at the higher baseband frequencies, thus increasing the signal/noise ratio at higher frequencies but reducing it at low frequencies. It also has the effect, in television systems, of reducing the interference levels with other microwave systems. The actual improvement in signal/ noise is approximately 2 dB, lower than the theoretical value by several dB, due to the problems of overmodulation and thus over-deviation.

The pre-emphasis circuit for television operates on the same principle but has different characteristics for PAL and SECAM systems. The circuit will improve signal/noise ratio, decrease the differential distortions in the TV signal as well as giving the operator the choice of lower bandwidth requirement or satellite power.

Demodulators

The demodulation process must recover the transmitted baseband signal without distortion in the case of analogue systems or in the introduction of errors for digital services. In practical terms this is going to be a compromise involving considerations of technology available, equipment size and cost.

The main method of modulation used in FDM systems is frequency modulation, as discussed in chapter 3. For FM systems the earth station designer has to consider two demodulator designs: first a broadband unit that can demodulate FM signals where the number of channels involved is greater than 250, and second a unit that not only demodulates the signal but also provides a measure of threshold extension and thus improved performance.

The wideband demodulator, in INTELSAT systems, can process up to 1800 channels and can be realised by a conventional form of FM discriminator as shown in Fig. 4.22.

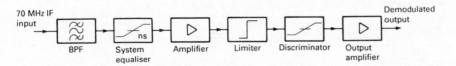

Fig. 4.22 Wideband FM demodulator.

The receiver usually operates at an IF frequency of 70/140 MHz and is restricted in bandwidth by a suitable band pass filter, equalised for group delay and amplitude before being applied to a conventional limiter/discriminator. The discriminator must be as linear as possible to prevent distortion and the output of the discriminator is buffered by a baseband processing system which not only de-emphasises the demodulated signal but also extracts the transmitted pilot which is transmitted between earth stations as an indication of circuit continuity. In addition the noise on the baseband signal is monitored to provide an indication of high noise and so allow the system redundancy system to operate, or for the system to be muted.

The threshold of the FM system has been defined in chapter 3 and this represents the limit of the input carrier to noise that a conventional FM discriminator can achieve. There has been a considerable amount of effort, invested over many years, to find methods of extending this threshold below the conventional limit and so improve system performance, allowing the system designer to make trade-offs between antenna size, HPA power and LNA noise temperature. The threshold extension demodulator can be in two basic forms:

Phase lock loop systems
This system is probably the most commonly used threshold extension system, whether used in a digital or analogue transmission system. The principle of operation is based upon the fact that the effective dynamic bandwidth of the circuit, as seen by the incoming modulated signal, is reduced, compared to the expected signal bandwidth necessary to fulfil Carson's Rule, which was set out in chapter 3.

The phase lock loop's effective bandwidth is narrower than the Carson bandwidth thus giving a noise reduction on the signal and therefore an improved signal/noise ratio compared to the theoretical. There are limits to the amount of extension that can be achieved due to the bounds put on the dynamic performance of the demodulator. The basic block diagram is shown in Fig. 4.23.

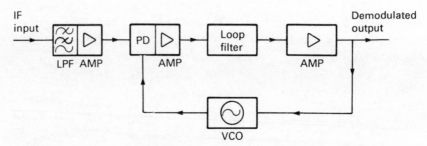

Fig. 4.23 Phase lock loop demodulator.

The IF input is applied, via filter amplifier to a mixer, which acts as a detector. When the phase lock loop is locked on to the input signal the VCO output frequency is equal to the input frequency with the error voltage being generated as a function of the phase difference between input and VCO signals. The loop filter determines the loop bandwidth which in turn sets the dynamic bandwidth of the PLL. This bandwidth varies with channel capacity and the loop filter is optimised for each capacity.

Information loss
The major problem associated with phase lock loops is the time taken to lock onto the input signal means a loss of information. This in itself is not important once the system is locked on but is very critical if the circuit loses lock and has to re-establish it.

Frequency modulated feedback systems
The basic FMFB circuit has a form as shown in Fig. 4.24.

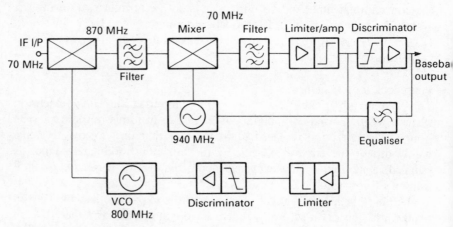

Fig. 4.24 FMFB demodulator.

The FMFB system obtains the necessary bandwidth reduction by the use of a narrow band filter placed in front of the FM discriminator. The detector loop takes the output of the discriminator and feeds it to the VCO which is acting as the local oscillator for the second mixer. This has the effect of reducing the FM deviation of the signal applied to the narrow band filter and it can thus pass through it without loss of information.

The nature of the system is such that it is necessary to keep the 70 MHz carrier frequency at the centre of the narrow band filter and therefore the effects of input frequency errors and oscillator drifts must be removed. This is done by the use of a second loop which controls the VCO that is acting as the local oscillator for the first mixer.

The circuit not only reduces the FM deviation but also the noise component of the signal in the same degree and thus the signal to noise ratio remains the same. The circuitry needs to be carefully optimised and the phase condition of the demodulator is adjusted by optimisation of the loop equaliser. This form of threshold extension demodulator is not used as much as the phase lock loop due to the increased circuit complexity.

4.3.5   Earth station control systems

Introduction

The earth station performance must be closely monitored and controlled, not only to ensure that the sub-systems perform correctly but also to allow the system to be reconfigured to cater for traffic changes and expansion.

The earth station operator has two forms of control and monitoring:

(a) Local level monitoring of sub-system performance, with the ability to take local control of the equipment;
(b) Remote monitoring and control, often at a central building where a number of antennas are involved.

The earth station operator must know the functional status of their system on a continuous 24-hour basis. There are three types of information that must be collated:

Alarms
These indicate either degradation or loss of performance, at unit or sub-system level. The earth station alarm philosophy is specific to each operator and depends upon the amount of information desired. The alarms can be situated:

(a) At individual plug-in unit level, which will give indications of the failure at board level of such parameters as power supply, signal input level, loss of lock, oscillator fail, etc.

(b) At equipment or cabinet level, where the unit alarms are gathered together at cabinet level and displayed locally.

(c) Area alarms which gather together all the unit and cabinet alarms and provide a display visible to the operators in that area.

(d) Centrally, on a computer screen or a large display panel giving immediate notice to the operational staff.

Supervisory information

This indicates the status of units or sub-systems. In general this information is of a passive nature, providing the operator with information on HPA status, antenna position, LNA performance, receive and transmit chain status as well as sub-systems such as power supplies, security systems, etc.

Control information

This shows the status of those sub-systems, that are set up in a redundant configuration, such as HPAs, LNAs, receives, transmit drives etc.

Computerised monitoring and control

The basic form of a control system is shown in Fig. 4.25.

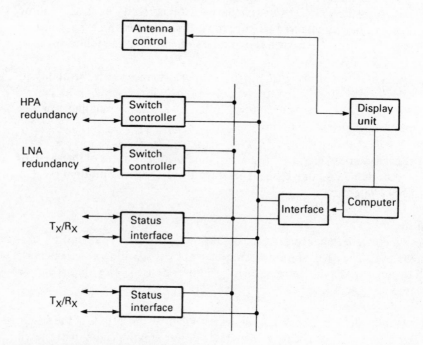

Fig. 4.25 Earth station controller system.

The central computer provides two major functions. First, it displays the current system connections and allows the operator to see the redundant configurations in detail, as well as where any faults are occurring. Secondly it communicates to a number of control shelves, via some form of interface bus such as RS232C, to not only monitor sub-systems status but also to initiate changeover procedures should any communication chain fail.

The computer stores, in its database, a complete record of system configurations, waveguide switch positions, etc. Status information is fed back to the computer from microprocessor controlled units which monitor tellbacks from the individual equipment. This information allows the central computer to control the interface units and commend them to change over switches, etc.

The advantage of the computer system is that it allows the operator to monitor at a variety of levels, from complete system down to individual unit, so simplifying not only the control of the system but also the maintenance of the earth station.

4.3.6 Terrestrial interfaces

The interface circuits between the earth station and the country's terrestrial system depend upon the type of service being provided by the system, whether it be analogue or digital and the method of terrestrial communications. In order to cater for the various interconnections a number of alternative equipments are available. The development of communications has moved from analogue to digital forms and the problem that has arisen is how to interface between new digital and existing analogue systems. The two equipments designed for this purpose are hypergroup codec and transmultiplexer.

Hypergroup codec

The level in the hierarchy at which interconnection takes place depends upon the terrestrial network and the form of data from the satellite earth station. In this particular instance it is at hypergroup level, which consists of 15 supergroups, giving a maximum aggregation of 68 Mbits at the output of the codec.

A hypergroup signal occupies from 312 kHz to 4028 kHz of the transmission band and in practice the encoding and decoding is done by frequency translating the input down to the 60 kHz to 3776 kHz band. A typical form of the encoder is shown in Fig. 4.26, with the decoder being the inverse. The A/D converters are fed via an AGC amplifier which operates on a pilot tone in the hypergroup signal to ensure correct system operation regardless of traffic loading. The ten-bit A/D converter quantises the signal which is then compressed to reduce the quantisation noise at low levels. The multiplexer takes the A/D output and multiplexes it to produce a 68.736 Mbit data stream. The composite bit stream is then coded into a continuous mark inversion (CMI) form, which is compatible with the 68 Mbit port of a 140 Mbit multiplex.

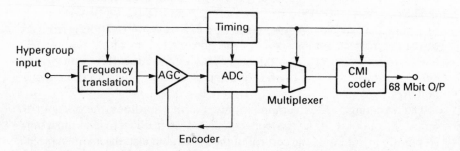

Fig. 4.26 Hypergroup codec encoder.

The decoder is the reverse of the above process with the 68 Mbit signal being fed directly into the CMI decoder, where the clock is recovered and used to decode the high bit rate signal which is then structured into 317 nine-bit words plus frame alignment and signalling. The digital signal is applied to the demultiplexer and then converted from digital to analogue, which is controlled in level, being followed by frequency translation back to hypergroup level.

Transmultiplexers

The transmultiplexer acts as a converter between a 60-channel FDM super-group and two 2048 kbit PCM data streams. The basic supergroup is in the 312 to 552 kHz frequency band, with the PCM bit streams carrying 30, 64 kbit channels.

The signal processing is done in a digital form in such a way that the conversion is carried out without the need to demultiplex the composite signal, thus reducing any degradation from the reprocessing itself. The conversion process applies to the PCM–FDM conversion (and vice-versa) but the transmultiplexer must take into account the signalling information contained within the signal. In the case of in-band signalling the transmultiplexer appears to be transparent to the signalling which passes directly through. If the signalling is Type R2 then a sub-system is provided that will convert the FDM out of band signal and the channel associated signalling in time slot 16 of the PCM signal.

4.3.7 Echo cancellation

One major problem with any communication system is the production of echoes. In terrestrial circuits the problem occurs when the telephone is connected by local line, which is usually a four-wire system. If these two systems are perfectly matched then no echoes occur, however in practice this is not possible and a return loss of 12 dB is common. The amount of echo effect is dependent on distance and for over 1000 miles is controlled by the uses of echo suppressors, which insert loss into the transmission path. This method has two major problems: first there is a limit to the insertion loss as this affects the signal as well as the echo and secondly, distortions are introduced, especially when both ends talk simultaneously in which case the suppression actually blocks the speech. This effect can be removed but the compensation itself can introduce echo bursts.

The problem of echo is considerably aggravated by the use of satellite communications. In this case the transmission delay is 260 ms which has to be added to the terrestrial links delay, giving a total in the order of 290 ms, necessitating the use of echo control. The most effective form of control in this case is the use of echo cancellation, which is superior to suppression.

The basic principle of echo cancellation can be understood by consideration of Fig. 4.27.

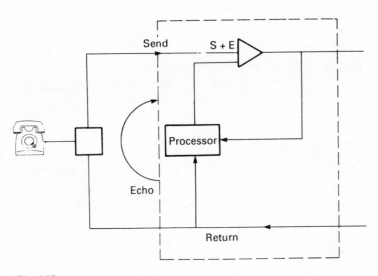

Fig. 4.27

The echo canceller is inserted into the line system at the same point as the echo suppressor. The signal is sampled and fed to digital circuitry that in effect produces a simulation of the actual echo signal. The echo model is then fed to an amplifier so that the real echo is cancelled. This particular system does not have the problems associated with echo suppression in that it can be configured to produce zero loss in the transmission path.

In practice the echo canceller is removed from the circuit while a both-talk situation is under way in order to avoid speech distortion. It is only used to produce an effective improvement in the circuit return loss performance, sufficient to give good speech tran...nission.

The problems associated with echo cancellers arise out of the transmission conditions and canceller implementation. The amount of echo to be cancelled will determine the required echo model storage. The response time of the system is important in determining how well the canceller responds to change in the echo path and produces echo return loss improvement.

Chapter 5

Satellite Communications for Telephony using INTELSAT

5.1 INTRODUCTION

The majority of satellite communication systems are used for transmitting telephone information, the other forms of communication, such as data and television, being of lesser importance in the profitable functioning of the overall system. The most important satellite system, INTELSAT, is the most developed with regard to the use of telephony and therefore this chapter will consider the INTELSAT system in order to give an overall view of the various types of service available. The development of INTELSAT has been detailed in chapter 2.

5.2 ORIGINAL STANDARD A EARTH STATION

This type of earth station is the major element in the worldwide INTELSAT network, operating in C Band (4/6 GHz) and conveying trunk telephony and occasional TV transmissions. The original specification defined the earth station, in terms of figure of merit, etc. and the fundamental parameter of the Standard A earth station is:

$$\frac{G}{T} = 40.7 + 20 \log \left(\frac{f}{4}\right) \text{ dB/}°\text{K} \tag{5.1}$$

The frequency, f, is the receive frequency measured in GHz. The other parameters set by INTELSAT were intended to define the earth station to be used by earth station operators. These parameters were both recommended and mandatory; the recommended characteristics were set to provide the most effective use of the satellite and the earth station design; the mandatory elements were those that had to be met by earth station operators before they were allowed to access the satellite. The mandatory parameters are:

5.2.1 Mandatory parameters

Figure of merit

It is essential that this figure be met to make best use of the satellite output power. This figure had to be achieved at all antenna elevation angles above 5° relative to the horizon.

The Standard A earth station figure of merit was revised in 1986 to take account of new satellite technology and the terrestrial developments associated with the siting of the earth stations closer to business premises, thus obviating the need for a terrestrial carrier link. The G/T was originally set at:

$$\frac{G}{T} = 40.7 + 20 \log \left(\frac{f}{4}\right) \text{ dB/°K} \qquad (5.2)$$

with the new specification being improved to:

$$\frac{G}{T} = 35.0 + 20 \log \left(\frac{f}{4}\right) \text{ dB/°K} \qquad (5.3)$$

Spurious radiation

It is important that the spurious radiation, in the form of intermodulation distortion, be restricted to prevent the spurious frequency components, within the transmission band, from interfering with the carriers transmitted from the other earth stations within the system. The limit of spurious radiation is defined as a power level in a 4 kHz telephony carrier bandwidth. This level is set at 26 dBW for the Standard A earth station.

Transmitter power stability

The power transmitted to the satellite has to be strictly controlled to within ± 0.5 dB of the nominal eirp. This stability, in terms of the overall system has to take into account not only the transmitter HPA stability but also the antenna structure stability and pointing error. The power received at the satellite must be controlled in order to ensure that there is no increase in intermodulation noise within the satellite transponder.

Transmit system characteristics

The transmission characteristics are set by filters which control the gain/frequency response and group delay characteristics of the signal, again in order to limit residual AM and thus the satellite transponder intermodulation noise. A typical form is shown in chapter 4.

Antenna characteristics

The antenna is defined in terms of:

Transmit sidelobe performance
This is set to within certain defined parameters which, in the case of the Standard A earth station, depend upon whether it was built before or after 12 March 1986 when the G/T parameter was revised.

The transmit sidelobe performance was in fact tightened and the changes are shown below: for $\dfrac{G}{T}$ of 40.7.

$$Gain = 32 - 25 \log \theta \text{ dBi}, \quad \text{where } \theta < 48° > 1°$$
$$Gain = -10 \text{ dBi} \quad \text{for } \theta > 48° \tag{5.4}$$

The gain G is defined as the gain of the sidelobes relative to that of an isotropic antenna that is transmitting in the direction of the geostationary satellite. The angle θ is measured relative to the main beam axis.

The new Standard A characteristics have a tighter specification, as below:

$$
\begin{aligned}
G &= 29 - 25 \log \theta \text{ dBi} & 1° &< \theta < 20° \\
G &= -3.5 \text{ dBi} & 20° &< \theta < 26.3° \\
G &= 32 - 25 \log \theta \text{ dBi} & 26.3° &< \theta < 48° \\
G &= -10 \text{ dBi} & \theta &> 48°
\end{aligned}
\tag{5.5}
$$

These characteristics allow the operator to exceed the sideband envelope for no more than 10% of the sidelobe peaks in co-polarised and cross polarisation senses.

Receive sidelobes
The sidelobe performance of the antenna has to be defined, not only for the reason that it affects the satellite transponder but that it has to be protected from external, terrestrial, interfering signals. The Standard A is therefore specified to the same level as the transmit performance, though it is not mandatory on the earth station operator.

Polarisation/axial ratio
The INTELSAT satellites transmit on one polarisation and receive on the opposite one. The actual polarisations used are defined by IESS 201 (Rev) and vary between IS 4A and the rest of the series, IS5/5A/6. The axial ratio must be strictly limited in order to use dual polarisation systems and a design goal of 1.06:1 has been set by INTELSAT. It has also become a design feature, for earth stations that are situated in high rain areas and therefore subject to the depolarisation effects of rain, to fit a depolarisation compensator which would automatically cut in when the actual polarisation is 5 dB

worse than the clear sky case. This parameter is not mandatory at C Band but may well become so at Ku and Ka Bands.

5.3 NEW STANDARD A EARTH STATION SPECIFICATION

5.3.1 General

The new specification has already been mentioned but it is worth considering the system in more detail. The original earth stations were specified in the early 1960s and had to take account of the technology available at that time, particularly the LNA front end, which initially had to use cryogenic techniques to obtain the required low noise figure. The mechanical technology available to the antenna designer was well proven, though not amenable to rapid improvement in the design of antennas to give smaller, more stable and lighter structures. The determining factor given the LNA technology, was the antenna reflector size and efficiency and in the beginning 29-metre reflectors were used with gains of 63 dB, with later increases to a 32-metre reflector size giving a consequent increase in gain to 64 dB. However the early earth stations utilised massive concrete structures as king posts, on which the antenna reflectors were mounted, with reflectors constructed of a large number of very accurately manufactured aluminium panels of high surface accuracy. The whole reflector had to be adjustable to an overall surface accuracy of 1.02 mm, all of which required a considerable steel backing to make the parabolic reflector construction sufficient to meet the pointing accuracy and tracking performance requirements. At the same time the whole system has to be moved in order to track the satellite. With this form of very large mechanical structure very large powerful electric motors were required to drive the antenna in azimuth and elevation and accurate and large bearings had to be used for mounting the whole antenna structure. The cost of such a structure and the whole earth station was considerable and once the civil works are included then the whole station would cost in the region of £5 million (1988 prices).

It was with this in mind that INTELSAT began a review of the Standard A requirements with the intention of reducing the figure of merit of the earth station. This became a possibility because of the improvement in satellite output power from the later generations of satellites. This modification would reduce antenna size and take account of the advances in the development of LNA technology to give an equivalent performance at a lower figure of merit. In addition to the possible size and cost reductions there were other reasons for a review. In general satellite services were being provided for the more specialised areas of business and rural communications where the smaller earth stations were co-sited with the user and therefore could take full advantage of one of the main characteristics of a satellite system, i.e. its point to multi-point capability which removes the need for a terrestrial network: a network which is costly to install, maintain and takes a considerable time to implement; satellites overcome all this. The early Standard A earth stations, due to their size and need for a relatively interference-free environment, had to be sited away from large centres of population on isolated sites, thus requiring a high capacity terrestrial link to carry the signals to the national network distribution centre.

A further advantage of the smaller earth station is that it can be used as an offload facility, where the main 32-metre Standard A operates as a primary path satellite while excess load would be transmitted to the satellite using the smaller Standard A. This would produce a more flexible, less costly network that can be used for a wide variety of services such as telephony, TV, data and TDMA.

5.3.2 New Standard A parameters

The figure of merit has already been listed, showing a reduction of 5.7 dB; this reduces the effective channel capacity by approximately 25%, depending upon the carrier to interference (C/I) that has been assumed for FDM/FM systems. The figure of merit can be realised in a variety of configurations, depending upon the trade-off made between antenna size and noise temperature.

A typical design would be as below:

Noise temperature	43 °K
System noise	13 °K
Antenna noise temperature	32 °K at 10° elevation
Total	88 °K
Overall system noise	19.5 dB/K
Antenna gain required	54.5 dB
Feeder/mismatch loss	0.5 dB
Overall antenna gain	55.0 dB

This means that the new Standard A antenna size could be reduced from 32 metres to 15.7 metres at an elevation angle of 10°.

eirp requirements

The eirp has to be considered for the various services that have to be provided such as telephony, TDMA and TV; this has been calculated for an HPA power of 3 kW or 70 dBW, a 972 ch/36 MHz FDM/FM carrier, a

17.5 MHz FM/TV carrier and a 120 Mbit TDMA carrier and the results are
shown in Table 5.1.

Table 5.1   eirp system margin

Ae dia m	FDM/FM	System Margin dB FM/TV	TDMA
16	− 0.6	− 0.7	− 5.7
17	0.0	− 0.1	− 5.1
18	0.4	0.3	− 4.7
19	0.9	0.8	− 4.2
20	1.4	1.3	− 3.7
21	1.8	1.7	− 3.3

It can be seen that the transmit requirements for FDM/FM and FM/TV
can be met by the use of an 18-metre antenna but it will not be feasible for
TDMA operation and this can only be achieved by a reduction in performance capability.

5.4   STANDARD A EARTH STATION REQUIREMENTS

5.4.1   Received power flux density

The amount of power received at the Standard A earth station is defined for
each satellite and polarisation, and is shown in Table 5.2.

**Table 5.2   INTELSAT Standard A received
power flux density**

Satellite	RPFD dBW/m^2 A Pol	B Pol
4A	− 124.3	–
5	− 130.2	− 132.0
5A	− 122.4	− 122.4
6	− 123.5	− 124.0

These transmissions are made at C Band in the frequency band covering
3700 to 4700 MHz as a mandatory requirement, although it is envisaged that
new earth stations should extend the lower frequency down to 3625 MHz.
Earth station planners take this into account in setting the earth station
parameters, even though the requirement is not mandatory, to ensure that
systems will work in the future regulatory environment.

5.4.2 Receive input linearity

The low noise amplifier front end to the earth station receive system is not supposed to add intermodulation noise to the received signal and INTELSAT define this as being less than 50 pWOp under the following conditions:

- Two carriers being received;
- Each received carrier is at least 3 dB below the expected receive power;
- The third order intermodulation of each carrier is 51 dB below that carrier level.

5.4.3 Telephone circuit quality

The overall quality of a telephone channel is defined by the amount of noise in that channel; a noise which is due to the satellite performance, the effects of the transmission medium between satellite and earth station and the earth station itself. The total allowable noise is 10 000 picowatts and is related to a zero level which is designated as pWOp, thus each noise degradation is in units of pWOp. This allowance of 10 000 pWOp is allocated in the following manner as in Table 5.3.

Table 5.3   Noise allocations for satellite link

Noise source	Noise level pWOp
Downpath thermal noise	4250
Up path thermal noise	1400
Transponder intermodulation	2350*
Terrestrial interference	1000
Earth station noise	1000

*Includes 500 pWOp of earth station HPA intermodulation noise.

5.4.4 Tracking system

Each satellite transmits a number of beacon frequencies to allow the earth station to track the satellite, which is kept in its position to within 0.1° in both N/S and E/W directions and takes out the movement and changes of the earth station antenna.

The beacon frequencies used are listed in Table 5.4.

Although these beacon signals are available on the satellite they are not all transmitted or used at the same time. It is possible to use other methods of

Table 5.4   INTELSAT Beacon frequencies

Satellite	Frequency MHz
4A/5	3947.5/3952.5
5A/6	3947.5/3948.0/3952.0/3952.5

tracking such as computer control and in this case the earth station can obtain station keeping information from the INTELSAT Control Center.

5.5   STANDARD B EARTH STATIONS

The Standard A earth station was intended for use on trunk communication routes, with a capacity between 24 and 1872 channels. These capacities were suitable for major users but for PTTs, that required only thin route gateway capability, this Standard A system was too large and the INTELSAT system provides for these users with a different standard, defined in IESS 202 as Standard B, which had a figure of merit as below:

$$G = 31.7 + 20 \ log \left(\frac{f}{4}\right) \ \mathrm{dB/^{\circ}K} \tag{5.6}$$

where the frequency f is in GHz and is in the band 3700–4200 MHz.

The trunk systems, employing relatively large antennas, have a tariff structure that makes the earth station user whose requirement is for thin route applications, operate in a less cost-effective manner. The INTELSAT Standard B makes use of a single channel per carrier (SCPC) system which allows the operator to transmit up to 60 channels. However if the user goes above this capacity then the tariffing encourages the operator to install a Standard A for those routes, while the Standard B is retained for thin route or off-load applications.

The growth of worldwide capacity requirements and the trend towards digital communications brought about the implementation of a TDMA system designed to give an efficient, high capacity system which would be used by the major operators utilising a channel capacity greater than 700 channels. In the event the initial take up of this service was not as great as expected due to the cost of TDMA earth stations and their operation and maintenance. These costs related to the fact that the individual earth stations used complex digital technology operating at 120 Mbits, and had special setting up requirements as well as the need to use highly trained staff, able to both operate and maintain the overall system. The thin route user has a number of ways in which they can increase their effective channel capacity, without the need to radically alter their earth station configurations. These are by companding and analogue circuit multiplication equipment (ACME), although ACME cannot be used at the same time as companding.

The Standard B can be realised in practical terms with an 11/12-metre parabolic antenna and Figure 4.12 shows a detailed design of the overall system. The basic elements of the receive system are outlined in chapter 4 and take account of connections to the LNA from the antenna, which in the case of the Standard B is direct from the antenna feed system by waveguide 11A, which has a loss of approximately 0.04 dB. The antenna is small enough to allow it to use a torque post mounting system and the connection from the LNA is via a cable banding system and a cross-site cable having a total loss of approximately 3.0 dB. Once the received signal is in the receiver area the signal is split down into separate channels for further processing.

5.5.1 Figure of merit

Effective antenna gain

Guaranteed gain (12 m antenna)	52.9 dBi
Feeder loss	− 0.4 dB
Mismatch loss	− 0.15 dB
Nett gain	52.7 dBi

System noise

Noise due to W/G loss/mismatch	9.7 °K
LNA/antenna mismatch	2.0 °K
Second stage contribution	1.0 °K
LNA/plate assembly	75.0 °K
Total noise	87.7 °K

Antenna noise temperature

Elevation	Noise temperature °K
10	32
20	23
25	22
40	20

Overall figure of merit

Total system noise (10° elevation)	20.8 dB/°K
Total system gain	52.7 dB
Figure of merit G/T (dB/°K)	31.9 dB/°K

The system performance required is 31.7 dB/°K and is therefore met by a 12-metre antenna and a 75 °K LNA. In practice this is the absolute minimum for the lowest elevation and the minimum antenna gain, which could 0.2 dB higher as an average figure.

5.5.2 eirp requirements

The eirp that is required depends upon the channel capacity to be supported. If it is assumed that the earth station is intended to carry up to 252 cfdm/fm channels and has to communicate between both Standard A and Standard B earth stations, a typical calculation has to take into account a number of factors which have been detailed in chapter 4, and are detailed below for a Standard B system.

Unconverter cable connection losses	− 4.4 dB
HP gain	+ 78.0 dB
Switch system loss	− 0.1 dB
W/G coupler loss	− 3.1 dB
Antenna W/G loss	− 1.1 dB
Antenna gain at 6175 MHz	+ 56.2 dB
Total gain	+ 125.5 dB

The eirp required is at a specified level of 78.8 dBW.

5.5.3 Out of band RF emissions

The use of SCPC means that there are no intermodulation products to take into account. The spurious signals are then made up of thermal noise and spurious tones. A typical Standard B performance for thermal noise is shown below:

$$F = F_1 + F_2 - \frac{1}{G_1} + L_1 F_3 - \frac{1}{G_1 \times G_2} \tag{5.7}$$

where F_1 = IF unit NF = 33 dB

F_2 = Upconverter gain = 13 dB

F_3 = HPA gain = 8 dB

L_1 = loss between U/C and HPA = 5.23 dB

G_1 = IF unit loss = $-$ 17 dB

G_2 = U/C gain = 15 dB

The above figures give an overall noise of 34.73 dB. If this is now translated into the amount of power in a 4 kHz bandwidth then:

$$T_{Neirp} = C_{eirp} - C + (F - 168) \text{ dBW/4 kHz} \tag{5.8}$$

where C = Input level to the IF unit dBW

This gives a thermal noise of $-$ 23.15 dBW/kHz. The spurious tones are dependent upon the upconverter spurious output, in practice less than $-$ 100 dBm, equivalent to a $-$ 8.88 dBW radiated output power and the spurious outputs from the HPA, usually $-$ 100 dB below the carrier output power. This gives a spurious output of $-$ 19.9 dBW.

5.5.4 Companded FM systems

The INTELSAT system has defined the circumstances in which companded FM can be used and in general these apply to Standard B communications between Standard A, Standard B, Standard C, and Standard F-3 earth stations. These guidelines do not allow the mixing of companded and non-companded services and are restricted to less than 252 channels between any two earth stations and less than 792 channels for any multi-destination carrier. The earth station user can also use partially companded carriers provided it is agreed with INTELSAT. In the INTELSAT document IESS 302 (Rev 3) the performance requirements and system parameters are set out in full; the IESS document not only prescribes the technical parameters of the system but also the actual characteristics of the compander itself.

The compander compresses the transmitted signal and expands the received signal. This compander is defined by various parameters which are listed in Table 5.5, in addition the method of using the compander is defined.

The new system had to be introduced in stages to take account of the introduction of the new INTELSAT Standard A performance specification. The basic parameters are:

Receiver performance

The earth station receiver has to perform a number of functions:

Table 5.5   Compander characteristics

Parameter	Value
Compression ratio	2 for I/P of − 60 dBmO to + 5 dBmO
	O/P of − 65 dBmO to + 5 dBmO
Unaffected level	− 10 dBmO ± 0.25 dB
Intermodulation	− 32 dB below single frequency level
Overshoot	< 15% with a 12 dB step change
Attack time	3 to 5 ms
Recovery time	13.5 to 22.5 ms
Compressor noise	− 45 dBmOp
Expander noise	− 80 dBmOp
Impedance	600 ohms balanced
Signal balance	> 60 dB over 300/3400 Hz (CCITT Rec 0–121)

(a) The demodulator system should utilise a threshold extension system. This is recommended on the grounds of the improved performance available, improved impulse noise performance, reduced interference from adjacent carriers, reduced AM/PM conversion noise and most of all, the improved overall system performance due to the raised threshold of the receiver. This improvement will be most evident at channel capacities up to 252 channels.

(b) Group delay and bandwidth characteristics are controlled by a filter in the receive chain and the INTELSAT specification details the required filter characteristics for the whole range of channel capacities, corresponding to bandwidths from 1.25 MHz to 25 MHz. These filters provide both linear and parabolic group delay equalisation of both the satellite and the earth station receive system. The receive filter has an amplitude/frequency response that not only defines the variation of amplitude over the bandwidth but also the out-of-band response, which reduces the adjacent channel interference to the required minimum limits. The design of these filters provides equipment designers with one of their most difficult challenges due to the tightness of the amplitude and group delay masks and the need to maintain the filter performance over a range of temperature and environmental conditions.

Transmit system characteristics

The required eirp characteristics are defined in IESS 302 for control of the performance between earth stations. The document defines the following:

(a) The baseband channels are assembled according to the CCIT system defining 12 and 60 channel groups. The ESC and energy dispersal signals are carried in the baseband below 12 kHz with the continuity pilot at 60 kHz. The load characteristics also conform to CCIT Recommendation G223, which sets the limits of baseband loading in order to prevent interference with other channels and intermodulation distortion.

The crucial parameter in this case is the rms multichannel deviation which must not be exceeded and therefore the organisation of the baseband must take account of the number of telephony and data channels in each group of channels. The mean level of the signal input into the baseband has to be controlled to within 1 dB of the specified figure by careful design of the baseband equipment to prevent over-deviation allied with an alarm system that will indicate over-deviation.

(b) Pre-emphasis is applied to the baseband signal as defined in CCIR Recommendation No 464-1 and the maximum baseband frequencies are defined in IESS 302 Table 6.

(c) RF energy dispersal is applied to the input signal into the modulator; the signal comprises a low frequency symmetrical triangular waveform. The energy dispersal signal has to have a level that will ensure that the maximum eirp in any 4 kHz band does not exceed the maximum eirp/4 kHz carrier by more than 2 dB. The INTELSAT document defines the ratio of unmodulated carrier to the maximum carrier power under full load conditions; the design also takes into account the fact that the low modulation carriers deviate from the standard formula due to their deviation from a true Gaussian distribution.

The triangular waveform has to be held to within ± 1.0 Hz, though it can be a frequency between 20 Hz to 150 Hz; it has to be combined with the baseband signal and attenuated above 4 kHz and its level is adjusted either continuously or as a stepped function.

Telephone channel quality

The Standard B earth station must conform to CCIR Recommendation 353–4 which sets the limits of noise and interference that are allowable in the design. This allows a total of 10 000 pWOp in any individual channel for good climatic conditions which hold for 80% of the month.

Companded FDM/FM carriers eirp

(a) eirp levels are defined for an elevation angle of 10° and assume that it is at the satellite antenna beam edge. This means that the parameters specified are geographically sensitive and should the earth station be situated in a different position then correction factors must be applied, correction factors that are calculated by INTELSAT and are detailed in IESS 402. The eirp levels are specified for the various carrier sizes up to 792 channels, also taking into account the satellite in use and whether the satellite is operating on a global, hemi or zone beam.

(b) eirp output levels have to be maintained to within ± 0.5 dB of the specified value. The factors affecting this stability are set out in chapter 4 and involve the designer in HPA design, antenna stability and tracking systems. The eirp can be allowed to fall below the nominal level by up to 2 dB, though this cannot be done without accepting that there will be poorer performance, which must be taken into account when communication is taking place between earth stations. In some cases the climatic conditions are such that it is inevitable that the eirp will fall below the nominal value, due to scintillation effects for instance. In this case the earth station should be fitted with some form of up-path power control. In order to control the up-path power it is not only necessary to sense the reduction in power by feedback from the satellite but also the circuitry must have a response time that will bring the eirp back to within ± 1 dB of its nominal value in less than two seconds.

eirp adjustment is necessary to allow the earth station operator to adjust the power level should INTELSAT request it – the range of adjustment being up to 15 dB.

(c) eirp limits off axis are defining according to the CCIR Recommendation 524–2 which sets the limits for both C and Ku Band systems. In addition the limits are set both for antennas built before 1 January 1989 and those built on or after that date. This tightening of the specification is to take account of the improved satellite station-keeping ability and the need to conserve the geostationary orbit. The limits are thus set at:

$$35 - 25 \log \theta \, dBW/4 \, kHz \, pre\text{-}1989 \qquad (5.9)$$

$$32 - 25 \log \theta \, dBW/4 \, kHz \, post\text{-}1989 \qquad (5.10)$$

$$39 - 25 \log \theta \, dBW/4 \, kHz \, for \, Ku \, Band, \, provisional/CCIR \quad (5.11)$$

The above parameters apply for elevation angles above 10° and are only to be exceeded if the earth station antenna can be shown to be better than the minimum specification.

Spurious emissions

The spurious emissions are limited to 4 dBW in any 4 kHz band, for any INTELSAT 5 and 6 satellites, at C and Ku Bands.

Carrier frequency stability

The variation in the transmitted frequency has two components; the initial setting of the frequency has a setting up error, due to test equipment and component tolerances and the various oscillators in the transmitter and IF chain have a frequency drift. To meet the INTELSAT requirements the frequency error must not exceed ± 150 kHz for all carriers above 5.0 MHz. For 2.5 and 5.0 MHz carriers the error is reduced to ± 80 kHz, while for the minimum carrier at 1.25 MHz the error must not exceed ± 40 kHz. The preceding figures are assumed to be long-term, i.e. one month, and the satellite retranslation frequency error is less than ± 25 kHz.

Gain/amplitude frequency response

The system has to maintain an amplitude frequency response to a mask as shown in Fig. 5.1. The restriction of the amplitude response is to control intelligible crosstalk and adjacent carrier interference. The crosstalk limit is set at a level of 58 dB below the carrier level.

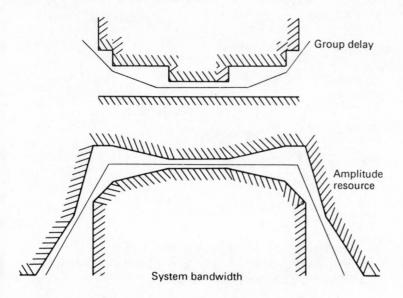

Group delay

Amplitude resource

System bandwidth

Notes: (1) Figures are symmetrical relative to centre frequency
(2) Figures are not drawn to scale
(3) Amplitude scale is linear in dB
(4) Frequency scale is linear in MHz

Fig. 5.1 Transmit amplitude and group delay requirements, earth station total.

The system also restricts the residual amplitude modulation, set at a level according to equation 5.12.

$$A_{mr} < -20(1 + \log f) \text{ dB relative to carrier level} \qquad (5.12)$$

The peak to peak modulation non-linearity must not exceed 1.5% for any channel capacity.

5.6 STANDARD C EARTH STATION

The previous gateway systems, using Standard A and Standard B earth stations, make use of C Band frequencies and INTELSAT have defined a gateway earth station for trunk, telephony and television that can operate at Ku Band frequencies. This is known as a Standard C earth station and its characteristics are defined in INTELSAT document IESS 203. The specification has, like the Standard A, been modified in the past few years to allow the earth station operator to use simpler and less costly earth stations. The specification is similar in form to the C Band systems but it has to take account of the special conditions that apply to Ku Band systems; this means that the IESS 203 document is set out for two different conditions.

5.6.1 Figure of merit

Clear sky conditions

The figure of merit for clear sky conditions is given by the following equation:

$$\frac{G}{T} = 41 + 20 \log \frac{f}{11.2} \, \text{dB/}^\circ\text{K} \tag{5.13}$$

$$\frac{G}{T} = 37 + 20 \log \frac{f}{11.2} \, \text{dB/}^\circ\text{K} \tag{5.14}$$

Equation 5.13 is the original value while equation 5.14 is the latest value. The figure of merit assumes that the atmospheric conditions conform to CCIR Reports 719 and 720, implying that there is very little rain or snowfall.

Degraded link conditions

The IESS document specifies a new G/T performance for degraded link conditions as in equation 5.15.

$$\frac{G}{T} = 37 + 20 \log \frac{f}{11.2} + X \, \text{dB/}^\circ\text{K} \tag{5.15}$$

where $X =$ the increased downlink degradation due to weather conditions; values are detailed in Table 5.6.

The degradation is calculated by INTELSAT for the specific satellite and earth station locations and not only takes account of the actual rain attenuation but also the increase in the receiver noise temperature. This distinction is important because the rain attenuation can be compensated for by simply

Table 5.6   Value *X* for Standard B satellite system

Satellite location	Satellite beam	Degradation	% Margin/year
325.5° to 341.5° E	West spot	13 dB	0.03
174.0° to 180° E	East spot	11 dB	0.01
307.0° to 310° E	West spot	13 dB	0.02
	East spot	11 dB	0.02

increasing the antenna size on a dB by dB increase but the noise temperature cannot be compensated for by a linear change of LNA noise temperature. The earth station designer has to take these factors into account when balancing the elements in the G/T trade-off.

The user must once again ensure that the earth station meets the IESS specification above an antenna elevation of 10°. For special local conditions the user must obtain special permission from INTELSAT.

5.6.2   Standard C antenna characteristics

The antenna performance that has to be met has been set to a much tighter specification for those systems installed after 1988. This improved specification applies to the sidelobe performance, which is detailed below.

Old Standard C antenna

Transmit sidelobes

$$G = 32 - 25 \log \theta \, \text{dBi} \quad \text{for } 1° < \theta < 48°$$
$$G = -10 \, \text{dBi} \qquad\qquad\qquad \theta > 48° \qquad\qquad (5.16)$$

where G is the sidelobe gain relative to an isotropic antenna
        θ is the angle between sidelobe and the mainbeam

Receive sidelobes

$$G = 32 - 25 \log \theta \, \text{dBi} \quad 1° < \theta < 48°$$
$$G = -10 \, \text{dBi} \qquad\qquad\qquad \theta > 48° \qquad\qquad (5.17)$$

The above parameters set the limit of the sidelobe radiation and that radiation must not exceed the limits by more than 10%, as detailed in CCIR Recommendation 580. The receive system parameters are not mandatory in this case but it is unlikely that these will be relaxed, in fact it is probable that the specification will be made tighter due to pressure on the orbit and the improvements being made to Ku Band antennas.

New Standard C antenna

Transmit/receive sidelobe performance
There is no difference between the receive and transmit parameters for the new antenna specification except that the receive performance is not mandatory. The sidelobe envelope is defined in detail as shown below:

$$
\begin{aligned}
G &= 29 - 25 \log \theta \text{ dBi} & 1° &< \theta < 20° \\
G &= -3.5 \text{ dBi} & 20° &< \theta < 26.3° \\
G &= 32 - 25 \log \theta \text{ dBi} & 26.3° &< \theta < 48° \\
G &= -10 \text{ dBi} & \theta &> 48°
\end{aligned}
\tag{5.18}
$$

Polarisation characteristics

The use of Ku Band makes it necessary to define the polarisation characteristics in order to prevent cross-polarisation effects and so be able to use different polarisation for separate frequency transmissions. For the INTELSAT 5 and 6 satellites the polarisations are linear and orthogonal for transmission and reception. The antenna axial ratio of transmission has to be better than 31.6 dB in the direction of the satellite and wherever possible the receive system should be better than this figure.

Antenna tracking performance

Satellite beacons
The INTELSAT 5 and 6 have two possible beacon frequencies, the second one being associated with the INTELSAT 5A (IBS) satellites.

(a) 11 GHz beacon frequencies: transmitted on a global beam on a right hand circularly polarised signal, at frequencies of 11.198 and 11.452 GHz.
(b) 12 GHz beacon frequencies: transmitted on a spot beam on a linearly polarised signal, at frequencies of 11.701 GHz for the 11.7–11.95 GHz band and at a frequency of 12.501 GHz for the 12.5–12.75 GHz band.

Station-keeping
The stability of the satellite is the major determinant in setting the tracking system parameters. The figures set out in Table 5.7 show the great improvement made between INTELSAT 5 and 6 satellites.

Table 5.7 INTELSAT satellite stability

	Stationkeeping	
Satellite	N/S	E/W
5, 5A and 5A 9 (IBS)	± 0.1°	± 0.1°
6	± 0.02°	± 0.06°

Tracking requirements
The INTELSAT document on Standard C earth stations does not provide mandatory directions to earth station operators regarding the methods of tracking but suggests that the system be made as flexible as possible even though this may increase the cost of the earth station.

Diversity systems
The Standard C system is unusual in that it allows for the use of more than one antenna to overcome the problems of operating at Ku Band in much more difficult environmental conditions. This method does require special system design in order to overcome the effects of the difference in the time that the two antenna output signals arrive at the combiner.

Received power flux density
The signal level received at the earth station is defined for the spot beam transmissions at 11 GHz and 12 GHz; these are detailed in Table 5.8.

Table 5.8   Standard C received power flux density

| Satellite | RPFD dBW/m^2 | |
	11 GHz Spot	12 GHz Spot
5	− 116.3	−
5A	− 112.3	−
5A (IBS)	− 112.3	− 116.8
6	− 110.4	−

Standard C earth station
The antenna most commonly used for the Standard C earth station is 18–19 m and with this it requires a low noise amplifier having a noise temperature of 150 °K. This is achieved by the use of thermoelectric-cooled FETs. The other parameter that has to be taken into account is the intermodulation noise contribution of the LNA, which is set at 50 pWOp in the IESS document. This will give a G/T or 35dB/°K.

5.7   C BAND VISTA SERVICES (STANDARD D)

The telephony earth stations, characterised by the preceding descriptions, have all been associated with the trunk routes and are used as gateway systems into the individual countries' telephone networks. The problem with these systems is that they do not address the needs of the small user in isolated circumstances, either rural communities, oil production platforms or disaster situations. To overcome this INTELSAT defined the conditions

necessary for a low capacity earth station to access the satellite; its formal
title was Standard D or VISTA service. The transmission allows the operator
to transmit up to 12 circuits, using the FDMA/SCPC mode of operation as
discussed in connection with the Standard B earth station.

5.7.1 VISTA system parameters

The VISTA service, as defined by INTELSAT, in IESS 204 and IESS 305,
makes use of global, zone and hemispheric beams to transmit the signals in
the 3700 MHz to 4200 MHz frequency band. The separation between the
SCPC/FM carriers is 30 kHz, apart from the two channels at the centre of
the frequency band, which use 45 kHz. The system is voice-activated in order
to conserve transmitter power. The figure of merit has two values to allow
the operator more freedom of choice.

Figure of merit G/T

The G/T is defined under clear sky conditions, with the antenna angle of
elevation greater than 10°.

$$\frac{G}{T} = 22.7 + 20 \log\left(\frac{f}{4}\right) \text{ dBW/°K} \quad \text{Standard D1} \qquad (5.19)$$

$$\frac{G}{T} = 31.7 + 20 \log\left(\frac{f}{4}\right) \text{ dBW/°K} \quad \text{Standard D2} \qquad (5.20)$$

Typical components for the earth stations are shown in Table 5.9.

Table 5.9   VISTA earth station design values

Element	Standard D1	Standard D2
Antenna	4.5 m	11 m
Tracking	Nil	Step
Gain G,	47 dBi	54 dBi
LNA	80 °K	80 °K

The antenna consists of a high efficiency Cassegrain system; the LNA
utilises GaAs FET amplifiers that require no cooling.

SCPC/CFM characteristics

The baseband characteristics define the channel parameters, IF bandwidth
and companding characteristics, as shown in Table 5.10. The frequency plan

set out by INTELSAT, in Fig. 5.2 is intended to allow the interleaving of the SCPC/CFM carriers in oppositely polarised satellite transponders, thus ensuring that the effects of intermodulation distortion are minimised. The frequency plan is also designed to ensure that the assignment of channels is very flexible, a flexibility that means that the SCPC/FM equipment has to generate carrier frequencies that are multiples of 15 kHz. The SCPC/FM equipment is also designed to be very flexible in terms of channel changes in order to prevent excessive disturbance to existing transmissions. There are also restrictions on the transmission of adjoining channels in order to meet off-axis emission parameters.

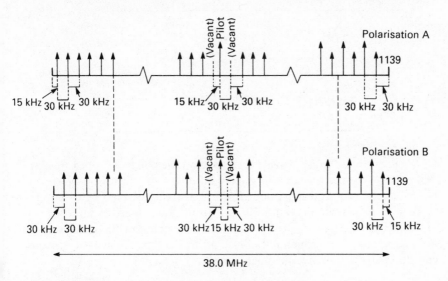

Fig. 5.2 SCPC/CFM frequency plan for full transponder operation.

eirp requirements for Standard D

The maximum value of eirp for the Standard D system is shown on Table 5.10.

Table 5.10   eirp/carrier for Standard D earth stations

Standard	eirp/carrier
D1	55.6 dBW
D2	51.7 dBW

The figures in Table 5.10 apply only to an antenna elevation angle of 30° and to take account of any changes in that angle a correction factor must be applied as shown in IESS 402. The eirp of the earth station must be able to handle up to 12 channels where the designer must multiply the channel eirp by the number of channels to be transmitted to give the total eirp. However for transmissions above 12 channels this linear relationship need not apply and the eirp can be reduced as shown in Table 5.11.

Table 5.11 Reduction of eirp as channel capacity increases

Number of channels	Activity factor (%)
> 12	100
12	85
18	72
24	67
30	64
42	60
60	57
> 60	40

The earth station must provide the user with the ability to adjust the eirp level in order to take account of local conditions. This adjustment must be a minimum of 15 dB and must not affect the stability of the output power, which must remain within + 1.0 dB/−1.5 dB for Standard D1 and ± 0.5 dB for Standard D2 earth stations. This takes no account of the losses due to rain precipitation etc. which may cause up to 2 dB greater power loss and will cause a reduction in system performance.

Baseband characteristics

The baseband parameters are shown in Table 5.12 which defines the system from the IF frequency downwards, including the carrier to noise parameters for the demodulator system. In addition the carrier and echo suppression have to be voice activated.

The compander must conform to CCITT Recommendation G-162 and the system must apply pre- and de-emphasis to each channel. The earth station must also receive a pilot generated by the satellite, to provide a means of controlling the carrier frequency and IF system gain.

Antenna requirements

The antenna is an important part of the earth station design and advantage can be taken of new design techniques to provide an improvement in terms of sidelobe performance. The specification is detailed below:

Table 5.12 Standard D earth station baseband parameters

Parameter	Value
Modulation	CFM
Companding	2:1 syllabic
Channel spacing	30 kHz
Channel bandwidth	30 kHz
IF noise bandwidth	25 kHz
rms tt deviation	5.1 kHz
C/N per channel at nominal	10.2 dB
C/No at nominal	54.2 dB/Hz
C/N at threshold	6.2 dB
C/No at threshold in IF BW	50.2 dB

Antenna gain
Usually the actual gain of the antenna sub-system is not specified by INTELSAT but in the case of a Standard D1 earth station the antenna gain is set to a minimum of:

$$G = 46.6 - 0.06(a - 30) \text{ dBi} \qquad (5.21)$$

where a = antenna elevation in degrees

The gain is set, in this case, because the system must not exceed the off-axis power transmission as specified in CCIR Recommendation 524-1.

Antenna sidelobes
The performance of the antenna to CCIR Recommendation 580 is made mandatory for transmit sidelobes:

$$G = 32 - 25 \log \theta \text{ dBi} \quad 1° < \theta < 48°$$
$$G = -10 \text{ dBi} \qquad\qquad \theta > 48° \qquad (5.22)$$

This applies to pointing angles greater than 1° off axis from the main beam, for smaller antennas i.e. $\dfrac{D}{\lambda} < 100$ the off-angle value becomes $\dfrac{100\lambda}{D}$. The earth station designer is encouraged to achieve an antenna performance of:

$$G = 29 - 25 \log \theta \text{ dBi} \qquad (5.23)$$

The receive sidelobes are again not mandatory but are restricted in order to reduce the effects of any outside interference and have the same specification as the mandatory transmit system.

Antenna axial ratio

The satellite system can be more cost-effective if dual polarisation is used. For the Standard D series of earth stations the antenna designer must meet an axial ratio of better than 1.06 for both D1 and D2 systems if the earth station is to be flexible enough for use over a wide range of geographical locations and able to operate on any of the satellite beams.

Antenna tracking

This is the same as for the Standard C system as it uses the same satellite, though this is applicable only to the larger Standard D1 earth station.

Received power flux density

The VISTA service operates on global, hemispheric and spot beam transmissions from the satellite. The earth station designer is provided with worst-case RPFD as in Table 5.13. The A polarisation (RHCP) is used on global, hemispheric and spot beams, while the B polarisation (LHCP) is used on global and spot beams.

Table 5.13   VISTA service received power flux density

Satellite	A–pol beams dBW/m^2	B–pol beams dBW/m^2
4A	-124.3	–
5	-130.2	-132.0
5A/B	-122.4	-122.4
6	-123.5	-124.0

EOW requirements

The order wire system is defined in IESS 403, with a recommendation that the Standard D earth stations be co-located with gateway systems such as Standard A and Standard B.

5.7.2   Earth station implementation

The basic parameters for the Standard D earth stations have been defined by INTELSAT as already described and these allow the designer to set each part of the earth station design. For this particular application the most important element in the earth station design is flexibility, not only in terms of the earth station equipment but also in considering the interconnection between the individual earth stations that make up the system which can be a star or meshed network.

The network architecture is set by the use to which the satellite system is being put; if the traffic requirements are relatively simple, with only a single master station being needed to route all the traffic requirements of the system, then a star system would be more suitable. The star system is the most cost-effective as an architecture though once the system grows too large then other earth stations, of a size equal to the main station, may have to be inserted into the star network.

The overall control of the system is often some form of assignment control, such as demand assigned multiple access (DAMA). In this case the satellite communication system would utilise a central control station. This station would monitor the frequencies in use, the management of the system interconnections, monitoring of the remote earth stations in the system and provide the operator with all of the necessary statistics associated with call connection and call charges.

The DAMA system works by answering a request for connection by allocating channels to the two stations. Signalling information is also transmitted when the call is set up and disconnected.

5.7.3   Typical Standard D channel capacities

The capacity available depends upon the satellite being used and the eirp used. In Table 5.14 an idea of channel capacity available is given. The satellite is assumed to be INTELSAT 6, with an eirp of 26.5 dBW for global operation and 31 dBW for zonal coverage.

Table 5.14   Standard D channel capacity

Standard	Satellite (global) 36 MHz	Satellite (zonal) 72 MHz
D2	1200	1368
D1	776	1368

5.8   INTELSAT STANDARD Z SERVICES

The INTELSAT organisation leases out transponders to individual countries to use as countrywide communication systems; this standard allows the operator to use a variety of antenna sizes, from 3 m up to 18 m in diameter, thus ensuring that they have a very flexible system which can be configured to provide a network of maximum efficiency and lowest cost. The great flexibility provided by the Standard Z has meant that in practice there are more of these in use than any other form of INTELSAT earth stations.

The Standard Z terminal is the most used of all the INTELSAT terminals, with over 70% of the installed bases following the Standard Z specification. The reason for this is that there are many countries that do not have the resources to install their own domestic system; in addition they do not have a developed telephony service and therefore the use of satellite transponders, leased from INTELSAT is an ideal solution to their problem, which can be fully operational in less than two years.

The specification follows a similar pattern to other INTELSAT leased transponder services in that the specification is very unstructured for:

(a) Maximum eirp per carrier;
(b) Modulation method;
(c) Figure of merit G/T;
(d) Transmit gain;
(e) Channel quality.

The user does have to fulfil certain conditions to avoid interference with other users, in addition they have to be flexible and meet certain mandatory conditions which are designed to protect the satellite.

5.8.1 Antenna system

Tracking capability

The need for tracking depends upon the size of the antenna and the frequency band in use, the Standard Z specification covering C, Ku and Ka Band. The earth station must be designed to operate within a particular arc for each ocean region as below:

Atlantic	307° E–359° E
Indian	57° E–66° E
Pacific	173° E–180° E

These limits are set to ensure that the earth station is designed in such a way that it is flexible enough in operation to move to another satellite if necessary. The station-keeping limits of the satellite have already been set out; a mandatory element in this specification is that the earth station has to be able to autotrack on the 5A, 5B (IBS) and 6. It must therefore be able to utilise any of the four beacon frequencies 3947.5, 3948, 3952 and 3952.5 GHz any two of which could be specified by INTELSAT.

Sidelobe performance

Design objective

For $\dfrac{D}{\lambda} > 150$ on existing antennas:

$$G = 29 - 25 \log \theta \text{ dBi} \tag{5.24}$$

All antennas must meet CCIR Rec 580–1

Transmit
It is mandatory that for pointing angles more than 1° off the main axis no more than 10% of the sidelobes can exceed:

$$\begin{aligned} G &= 32 - 25 \log \theta \text{ dBi} & 1° < \theta < 48° \\ &- 10 \text{ dBi} & > 48° \end{aligned} \tag{5.25}$$

Receive
The receive specification is as described for transmit but is not mandatory.

Polarisation

There are various mandatory requirements on the earth station operator relating to the way that polarisation affects the earth station design. The antenna has to be designed to operate on either of the circular polarisations that are orthogonal to one another, in addition it must be adjustable to allow it to match the satellite polarisation to within 1°.

The other parameter that is designed is axial ratio, a necessity on dual polarisation systems for each frequency band on both transmit and receive. The values vary between satellite but a design aim is 1:06.

5.9 SMALL STATION IMPLEMENTATION

5.9.1 Introduction

There have been a number of companies, across the world, that have responded to user requirements by designing a range of earth stations that will cover many of the INTELSAT specifications. The small earth station market especially has been addressed.

The systems designed for the VISTA specification use SCPC transmission, with FDMA access. The centralised control method can be either pre-assigned or demand assigned, the choice depending upon the system requirements, though the most cost-effective method is the DAMA form which uses less transponder time and is more flexible in operation.

The major use of the small telephony terminal is in a rural situation, a situation which is defined by CCITT which lists a number of characteristics that must be taken into account in any earth station design; these are:

- The community to be serviced should be either a village or small town, essentially scattered in nature;
- Isolated from other communities;
- Lack of facilities, such as electricity, from a mains source;
- Low skill level, in terms of modern technology;
- Low level of affluence;
- Poor geographical environment.

The above characteristics lead the system designer to a number of conclusions with regard to the earth station parameters, i.e.

(a) The terminal must be cost-efficient at small channel capacity.
(b) The terminals must be easy to transport and therefore be small in size and low in weight.
(c) Low power consumption is essential, with the ability to operate from solar power if at all possible.
(d) The maintenance of the terminals must be very simple. If the maintenance philosophy is one of unit replacement then the physical construction of the terminal must be simple to access and individual units easy to replace.
(e) The earth station must be low cost, both in terms of initial investment and the overall running costs.
(f) The terminal must be robust to take account of transportation over rough terrain.

5.9.2 Typical earth station

Antenna system

The antenna available for the earth station user has to be flexible in order to provide for differences in channel capacity. The general form of antenna is usually a Cassegrain system, giving as it does high gain and simplicity of construction.

The size of reflector will determine if tracking is required, with diameters to 6.5 m allowing fixed antenna operation. Above that diameter the use of simple step tracking is sufficient. This antenna would be 4.5 m for a Standard D1 terminal, giving a G/T of 23 dB/°K, assuming an antenna gain of 47.3 dB. For television operation it will be necessary to use a larger antenna of 6.5 m in order to provide enough eirp.

Low noise amplifier

The LNA must be compact and of as low technology as possible; in this case a GaAs FET transistor unit requires no cooling system, is small in size and low cost. A noise temperature of 80 °K is easily attainable with the ability to mount the unit directly onto the antenna feed.

Signal processing

The processing of the received and transmitted signal is usually designed in a modular form in which each SCPC channel is processed in a separate unit and it is therefore simple to add or modify the terminal as traffic requirements change. A block diagram of the system is shown in Fig. 5.3.

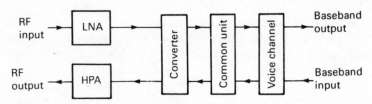

Fig. 5.3 VISTA signal processing system.

The common requirements, such as AGC and frequency synthesisers, are housed in a separate unit in order to reduce channel costs. The voice channel units contain threshold extension demodulators to improve system performance as well as companders which can give a further 15 dB. The theme of low cost is continued by the use of voice activation, which only uses the satellite channel during actual speech transmission. The unit would also contain echo suppression and the required terrestrial interfaces.

System access method

The access method is to some extent limited by the imposition of SCPC as the transmission mode and will obviously depend upon the interconnection mode between earth stations. In general the communication is two-way but is routed through a central hub in a star system. The DAMA system is usually housed in the central station within the system. That earth station has a large antenna to Standard B performance in order to give a high G/T which will reduce the power required from both the earth station and the satellite. The actual eirp to be transmitted cannot exceed 1 W per channel and is proportional to the satellite transponder gain and the earth station antenna gain; for the VISTA service an earth station will require 55.6 dBW per channel.

The overall network is often not a simple star but a mixture of star and mesh. This can be done by taking a number of local groupings as individual star systems with an overlay taking into account interconnections between the star networks. The DAMA system in general must provide the following facilities:

- It must manage the channels available;
- It must monitor the frequency of operation;
- It must manage the start up of and the cessation of calls;
- It must perform all the statistical operations necessary for the effective operation of a telephone system i.e. call billing and charging, fault monitoring and logging as well as any remote earth station monitoring.

5.10 SUMMARY

The use of satellite communications for telephony service is well served by a range of INTELSAT terminals, as well as private or regional systems, which over the years has led to the comprehensive provision of earth stations worldwide.

Chapter 6

Mobile Satellite Communications

6.1 INTRODUCTION

The use of satellites to provide a reliable communication system for maritime and aeronautical services is a relatively recent innovation and there has been a great increase, not only in the service availability, but also the number of services available. The major method of communication for maritime communications had previously been HF, which was used as a communication method by Marconi in 1899 when he transmitted messages between his yacht SS *St Paul* and the Isle of Wight. This method was developed over the years and has been installed in 98% of all ships, with many countries passing laws to force ship owners to use HF communications.

The problem of HF communications is its unreliability due to the nature of the ionosphere which varies both with the frequency of operation and the actual time of day. This variability of the ionosphere often results in loss of communication for days at a time which in maritime terms can produce serious problems for ship owners. The natural characteristics of the ionosphere have to a great extent been overcome by the skill of the equipment designer, not only in measurements of the ionosphere to predict its behaviour and so determine the maximum usable frequency at different geographical positions and times of the year, but also in devising systems that automatically change transmitters and receivers to the optimum frequency of operation. This process went so far that even though the death of HF was continuously predicted it has gone from strength to strength.

The coming of satellite communications opened up the possibility of its use as a means of communication from ship to ship, ship to shore and vice versa. It was seen to be capable of providing a non-fading reliable communication system, though it presented serious problems with regard to its ship earth station design.

The beginning of the worldwide maritime communication systems, namely INMARSAT, can be traced back to an international conference organised by the Inter-Governmental Maritime Consultative Organization (IMCO) in 1973. While the work following the IMCO conference was carried out, a practical system was launched by the Comsat organisation in 1976. This system was named MARISAT and was limited to the Atlantic

Ocean. It offered the user only two telephone channels and 22 telex channels, mainly because it shared its satellite with the US Navy and was thus limited in the satellite power available. This initial service was gradually expanded to give eight voice channels and continued its operation until 1982. The IMCO impetus led to the signing of the INMARSAT convention and operation agreement in 1976. The INMARSAT organisation is similar in form to the INTELSAT one in that countries, or their designated representatives sign the operating agreement and contribute financially to the operation of the organisation. The day to day operation of the system is controlled by a Directorate, headed by a Director-General, who is responsible to a Council, consisting of 18 of the major investors and four elected representatives (designed to protect the interests of developing countries and different areas of the world). The overall view of member nations is conveyed to the Council, via the Assembly, this consisting of one representative from each nation.

The MARISAT system was the foundation of INMARSAT, both in terms of the satellites used and the ship earth stations technical specification. This allowed a smooth transition to take place in the operation of the two systems. In 1982, in addition to the original satellites, INMARSAT leased a satellite from the European Space Agency (ESA), giving one satellite per Ocean Region. By 1984 the organisation had leased three maritime communication sub-systems on the INTELSAT 5 satellites which were used to provide cover for the Indian Ocean Region (IOR) and spares for the Atlantic and Indian Ocean regions. By 1986 the original MARISAT satellites had been replaced by an INTELSAT transponder, as shown in Table 6.1.

Table 6.1   INMARSAT satellite positions

Region	AOR (Spare)		IOR (Spare)		POR (Spare)	
Satellite	MARECS	A.IS5F6	IS5F5	IS5F7	MARECS B	MARISAT F3
Position	26° W	18.5° W	63° E	60° E	177.5° E	176.5° E

The orbital position of the satellites was obviously arranged to give optimum access from any point of the ocean region such that coast earth stations will operate above a 5° elevation angle (see Fig. 6.1).

6.2   INMARSAT OPERATIONAL SYSTEM

The system requires that the satellite can operate in two basic modes; the first operating at L Band between ship and satellite and the second from shore to satellite at C Band. The L Band transmission from the ship to the satellite is converted to the 4 GHz band before being transmitted back to the coast earth station. The complex process is shown in Fig. 6.2.

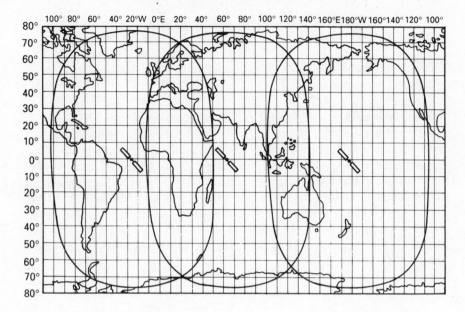

Fig. 6.1 INMARSAT global coverage.

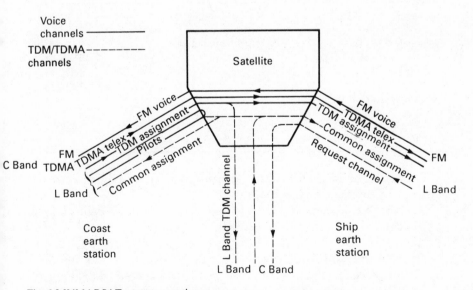

Fig. 6.2 INMARSAT system operation.

The INMARSAT system consists of three major elements: coast earth stations, network co-ordination stations and ship earth stations.

6.2.1 Coast earth stations

The onshore coast earth stations (CESs) operate within the INMARSAT system but are actually owned and operated by the PTTs within the countries of operation. The terminals act as a transmission medium between the terrestrial services such as telephone, telex, data etc. In addition to this function, the earth station acts as a control and access centre, it monitors the status of each channel, allocates channel frequencies and routes and connects calls between ship and terrestrial networks. Apart from these communication functions it logs calls and produces billing information to allow call charging. The original services provided by the INMARSAT system consisted of:

(a) *Telephony*: using companded, narrow band FM operating in an SCPS/FDMA mode;
(b) *Telex*: having 22 telex channels carried on a TDM carrier operating at a 1.2 kbit rate and using bi-phase coherent phase shift keying.

This service has now been expanded to include duplex data using NBFM.

Table 6.2   Coast earth station parameters

Parameter		C Band	L Band
Frequency band Rx MHz		4192–4200	1535–1542
	Tx MHz	6417–6425	1636–1644
Polarisation Rx		LHC	RHC
	Tx	RHC	RHC
Antenna gain dBi Rx		50.5	29.0
	Tx	54.0	29.5
G/T dB/°K		32	2.0
Transmit power dBW		66	36

The coast earth station has the performance capability set out in Table 6.2 and the CES format shown in Fig. 6.3.

The system has to operate in a dual-band mode, in C and L Band and this can be achieved by either of two methods:

(a) The antenna will be a single unit that can operate across the total frequency range required. In addition it has to operate in dual polarisation mode, meaning that any system will require an 11-metre diameter

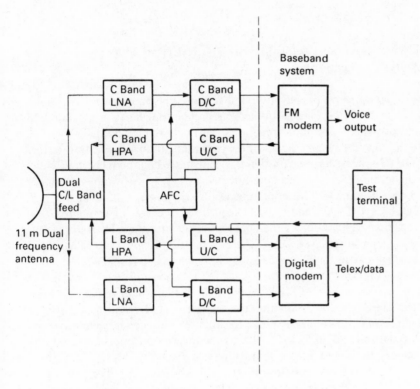

Fig. 6.3 CES format.

parabolic antenna with a dual-band feed that is in itself complex and expensive to produce.

(b) The CES can consist of two separate antennas, the first operating at C Band, with a relatively simple feed compared to the single antenna case and therefore is not expensive. The second antenna is of much smaller size, operating at L Band, with both antennas being coupled together for the purpose of tracking the satellite. This method was used for the first Scandinavian coast earth station, which was one of the first systems built specifically for the INMARSAT system.

The choice of method is entirely a matter for the operator of the CES and will be a decision based on the trade-off between cost of the system and the complexity of coupled antennas. Other factors that must be taken into account are associated with the electronic equipment and the additional land required for the two system antenna solution. A typical RF system is shown in Fig. 6.3.

The RF system comprises:

(a)   C Band receive and transmit chains;
(b)   L Band receive and transmit chains;
(c)   An AFC system.

The C Band receive system front end is a low noise amplifier with a noise temperature of 40 °K. The amplifiers are often connected in a redundant mode to provide the required reliability and the specified figure of merit of 32 dB/°K. In every system the maritime communication signal is transmitted on the LHCP, while the beacon signal is transmitted on the RHCP. In many cases this beacon signal is recovered by using a separate low noise amplifier system and beacon ground communication equipment (GCE) chain. It is possible to avoid this additional expense by the use of a polarisation injection filter which couples the beacon output into the LHCP path, thus using the main signal LNA before the beacon output is recovered and used for step track purposes. The price to pay for such a cost-saving is a reduction of 0.1 dB in the beacon signal level. The C Band downconverter converts the 4 GHz signal down to 70 MHz, using a double downconversion process, with the AFC module reducing the carrier signal frequency error to ± 20 kHz from a possible error, due to satellite frequency translation errors and Doppler shift, of ± 55 kHz. The downconverter output is fed to an FM demodulator to recover the voice transmission from the ship.

The C Band transmit system utilises 3 kW klystron amplifiers and operates with up to 60 channel carriers. The transmitter drivers are used to remove the effects of satellite and Doppler; this is achieved by an AFC module which receives a signal from the L Band receiver system which has a frequency variation due to satellite and Doppler effect. This signal is compared to a clean reference signal to produce a control signal that is used to pull the C Band carrier frequency an amount that is equal and opposite to the errors. The reason for the AFC system is that the ship earth station demodulators are narrow-band and cannot therefore tolerate large input frequency errors. The AFC system operates in such a way that the frequency of the signal received at the ship earth station is within ± 230 Hz relative to the centre frequency. It also incorporates a search system to look for the pilot signal should the system lose synchronisation.

The L Band sub-system is used to transmit telex messages and channel assignment instructions to the ship earth station using a 1.2 kbit TDM carrier. These signals are amplified using a 50 W linear amplifier. The receive system utilises a 220 °K low noise amplifier connected as a redundant system.

The access and control equipment interfaces with the RF system and brings together FM and digital modems, order wire equipment, telex switching and test terminals.

6.2.2 Network co-ordination station

An essential element in the overall system is the network co-ordination station (NCS). There is one NCS for each ocean region, the Atlantic Ocean is controlled from the earth station at Southbury in the USA; the Pacific Ocean is controlled from an earth station at Ibaraki in Japan and the Indian Ocean is also controlled from Japan, from an earth station at Namaguchi. The whole system is overseen from the INMARSAT headquarters in London. The NCS is assigned a unique TDM carrier frequency and this frequency is continuously monitored by the region's coast earth station and every ship earth station (SES) within the region. This monitoring continues on board ship during voice but not telex transmissions. The NES has a variety of co-ordinating functions:

(a) If a call is originated from an SES it sends a call request, using a 4.8 kbit random burst TDMA signal sent over the L Band request channel. The transmitted signal goes via the CES and the NCS where channel assignments are made from the NCS via the common assignment channel.

(b) If a call is from a terrestrial location it is routed via the CES which allocates two channel frequencies from its allocation and asks the NCS for an assignment. The function of the NCS is not only to assign a channel but also to tell the SES and CES that an allocation has been made, using the common assignment channel, thus allowing the call to be connected.

(c) If a telex request is made from the ship, via the L Band random access channel, it is received by the CES which allocates a channel and obtains permission from the NCS for use of the channel. If the telex message is routed from the terrestrial location, via the CES, then this also is approved by the NCS which informs the ship with which the link is being made, and the communication link is made over the CES TDM channel.

(d) To carry out its control function the NCS must store information on the status of the whole region's telephone channel usage to ensure that it is aware not only of the level of activity but also the origination and destination points of each call. This information not only allows it to control the region but provides traffic analysis information that can be used for forward planning purposes.

(e) In emergency situations the NCS can be used to break into any on-going calls to transmit SOS signals.

6.2.3 Ship earth station

The ship earth station presented designers with considerable problems, particularly with regard to the antenna stabilisation requirements. The antenna has to track the satellite but also take account of the direction in which the vessel is moving, as well as the roll, pitch and yaw characteristics of the vessel itself. The equipment design was complicated by the frequency band assigned to the communication links which produced a conflict between the need for very small lightweight terminals which would not affect the ship's stability or get in the way of existing antenna installations and yet would provide a wide range of communication services perhaps needing considerable bandwidth at very low cost.

The L Band technology available at the time of the initiation of the MARISAT services, whose specification carried on into INMARSAT terminals, meant that early terminals, especially with regard to below decks equipment, was bulky, heavy and very expensive. In fact it was a major obstacle in the way of the growth of the satellite service worldwide, a barrier compounded by the fact that the use of HF was mandatory in many countries and satellite communications could only be used at sea, not when the ship was berthed and therefore communication was severely restricted. The price barrier was, in many ways, the most important element in the marketing equation and fierce competition between manufacturers once the INMARSAT system was in place, forced down the price of equipment and led to new generations of ship earth station, that were smaller, lighter and more cost-effective. The INMARSAT organisation created a number of SES standards to cater for different requirements.

Standard A

The basic parameters of a typical ship earth station are:

Transmitting frequency	1636.5–1645.0 MHz
Receiving frequency	1535.0–1543.5 MHz
Channel capacity	339 (25 kHz spacing)
Figure of merit	-4 dB/°K
eirp	36 dBW
Antenna	90 cm parabolic reflector
Stabilisation	Gyro-stabilised/4 axes
Environment	-35 to $+55$ °C
	95% humidity
	100 knots wind speed
	Vibration 10–15 Hz peak amplitude 0.76 mm
Communications mode	NBFM with companding
Telex	4.8 kbits/1.2 kbits BPSK
Data	NBFM without companding

The basic SES package consists of an above-decks unit: a 0.9-metre parabolic antenna with sufficient gain to give a − 4 dB/°K figure of merit for the overall receiving system. The antenna could use a variety of methods for stabilisation but will almost certainly use a four-axes, gyroscopically stabilised system that is capable of automatically tracking the satellite. The antenna and feed are protected from the elements by a radome giving an overall above-decks diameter and height of approximately 1.4 metres. The design of the system is such that the SES will track the satellite under movement conditions as shown in Table 6.3.

Table 6.3   Ship earth station environmental parameters

Parameter	Value
Roll	± 30° for 8 sec
Pitch	± 10° for 6 sec
Yaw	± 80° for 50 sec
Surge	± 0.2 g
Sway	± 0.2 g
Heave	± 0.5 g
Turning rate	6 degrees/sec
Speed	30 knots

The radome is a vital element in the construction of the ship earth station in that the ship will encounter high humidity and temperature, sea spray and icing conditions that could put an inch of ice on the above-decks equipment. In addition the above-decks equipment must be able to operate within specification during wind conditions of up to 90 knots and considerable vibration due to the ship's operation.

The antenna is coupled straight through the deck to the below-deck unit. It also provides an output to the gyro-compass to facilitate tracking of the satellite. The below-decks equipment will consist of the RF package, LNA, HPA etc. and the processor equipment needed to make the operation simple and user-friendly.

The terminal operates at L Band, with the ability to tune to any one of the 339 channels, separated at a 25 kHz spacing. The power transmitted to the satellite is 36 dBW. This transmitted power is used to support a variety of services previously listed. For voice traffic the noise is muted between the speech by a voice activation system. The ship terminals have a wide range of communications methods, other than voice, such as FAX, system operation from a teleprinter or PC, as well as a synthesised voice that guides the operator through the necessary operational steps. In fact it is now possible to have completely unattended operation with the advent of advanced microprocessor techniques.

One of the crucial elements in the cost-use equation is the cost of installation, which means that due consideration must be given to the mechanical design of the system from the point of view of simplicity of installation and the need to provide a comprehensive worldwide maintenance structure to service the ship earth station installations.

Standard B earth station

The Standard A earth station is limited in its facilities and INMARSAT generated a new standard, Standard B, which gave the following services:

(a) Telephony using digital modulation, coding and speech processing to achieve the required quality.
(b) Low speed data services up to 9.6 kbits, which will support telex, teletext and facsimile.
(c) High data rate services requiring up to 16 kbits.

The earth station would have the characteristics shown in Table 6.4.

Table 6.4   Standard B ship earth station design parameters

Transmission rate	32 kbits
Channel bit rate	16 kbits
Modulation	4 PSK
G/T	$-4\,\mathrm{dB/^\circ K}$
Antenna gain	21 dBi
Coding	Adaptive predictive coding
FEC	1/2 rate convolutional
Decoding	Soft decision Viterbi
Telephony mode	Voice activated

The above system is intended to have the same performance as the Standard A earth station, though it is intended to reduce the figure of merit in the future terminals.

The link budget for the Standard B is shown in Table 6.5.

Standard C earth station

The Standard A ship earth station is designed to operate on ships above a particular size, about 10 000 tons. This severely restricts the ship earth station market and a new standard was designed to provide low cost, lightweight terminals that will give the user a digital capability of up to 600 bits/s. The terminal has the characteristics as shown in Table 6.6.

The system can operate in a redundancy mode in which the effective transmission rate is doubled to 1200 bits/second by sending the message twice.

Table 6.5   **Standard B ship earth station link budget**

Satellite/ship	
eirp	13 dBW
Free space loss	189 dB
G/T	− 4.0 dB/°K
C/No	49 dB/Hz
Overall C/No	46.6 dB/Hz
Ship/satellite	
eirp	31 dBW
Free space loss	189 dB
Satellite G/T	− 12.5 dB/°K
Up path C/No	58 dB/°K
Satellite C/IMo	69 dB/Hz

Table 6.6   **Standard C ship earth station design parameters**

Parameter	Value
Frequency	As Standard A
eirp	13 dBW
Modulation	BPSK
Figure of merit	− 22.8 dB/K
Coding	Viterbi rate 1/2
Interleaving	Selectable
Transmission rate	600/1200 bits selectable
Carrier/noise	36 dB/Hz
BER	10^{-3}

6.3 INMARSAT DEVELOPMENTS

6.3.1 Aeronautical

The latest frequency allocations appear in Fig. 6.4, which shows that the previous maritime allocations have been modified to allow part of the frequency band to be used by INMARSAT for aeronautical purposes.

The maritime frequency band extends from 1545 to 1559 MHz and 1626.5 to 1645.5 MHz. These frequency bands were originally intended solely for aeronautical applications but due to pressure on the spectrum they have been modified to give some of the band to land mobile services. The original allocations were made in 1971 with the WARC Mobile Conference in 1987 setting new allocations to take into account the new developments. These came into force in 1989, after a long period of trials using a variety of

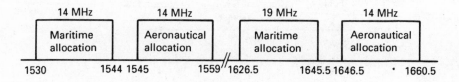

(a) Existing mobile satellite frequency spectrum

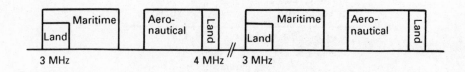

(b) New mobile satellite frequency spectrum

Fig. 6.4 Mobile satellite frequency allocations.

equipment and systems. The service was aimed at airline passengers, requiring a telephone service, airline operational arrangements and air traffic control purposes. The communication and physical requirements of an aeronautical system place considerable restraints on the equipment and system designers. The system implementation is shown in Fig. 6.5.

The system has three basic elements:

The INMARSAT satellites

These operate in the same frequency band as the maritime service. The link between the fixed earth station and the satellite uses a frequency in the 6/4 MHz band and L Band, while the link between the aircraft and the satellite takes place at L Band. The system utilises three satellites to give global coverage which can be INTELSAT 5, Marecs or INMARSAT series.

Aeronautical ground stations

These are the connection between the satellite and the terrestrial network. They are in fact modified coast earth stations, which allows them to operate in the aeronautical segment. This involves modifications to the earth station's up/downconverter input frequencies and the automatic frequency correction circuitry, which has to cater for considerably larger Doppler shifts.

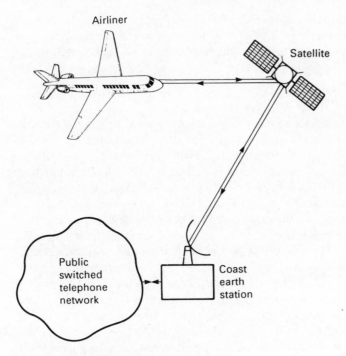

Fig. 6.5 Basic INMARSAT Aeronautical system concept.

Aircraft earth stations

These can be implemented in three configurations, corresponding to the various transmission capabilities of the earth station. These three options will allow the operator to have a choice that ranges from data only (up to 300 bits/sec), through voice only to voice and data capability which will give a 9600 bit/sec information rate.

System design considerations

The basic limitations placed upon the aeronautical system are:

(a) Limited bandwidth.
(b) Limited eirp due to the restrictions on both the aeronautical earth station's high power amplifier and the physical limitations of the earth station antenna.
(c) Doppler frequency effects due to the movement of the aircraft, which is much more significant for aeronautical than maritime systems.
(d) The effect of the multipath fading, which again is different in characteristics to the maritime system in that the fading bandwidth is larger by a

factor of 10 to 20 times. This increase in bandwidth has the effect of increasing the rate of received signal variation at frequencies up to 3–4 kHz, which places considerable constraints upon the system designer.

System implementation

The limitations listed above determine both the system design and the individual modules that go to make up the total aeronautical earth station. The limited bandwidth available points to frequency re-use with high power amplifiers using Class C mode of working to provide a very efficient conversion from the aircraft power supply to the RF output power. This means that the eirp can be achieved by the use of a 40 W amplifier for the voice/data operation.

The Doppler frequency effects impinge on the system design in areas such as forward error correction, coding, interleaving and increasing bandwidth. The Doppler shift has to be removed at the receiver and the earth station frequency offset has to be compensated for by offsets in the AES transmitter frequency.

The amount of multipath fading is dependent upon the elevation angle of the AES antenna; at elevation angles above 25° the effect is only second order and can be discounted. The effect of fading is controlled by a system of bit interleaving.

The above requirements have led to the parameters, detailed in Table 6.7, being set by INMARSAT.

Table 6.7

Parameter	Data only		Voice/Data
Information bit rate	< 300		9600
Transmission bit rate	600		21 000
Modulation	Symmetrical BPSK		Offset BPSK
Carrier spacing	2.5 kHz		17.5 kHz
G/T	− 26 dB/K		− 13 dB/K
FEC	$\frac{1}{2}$ rate Viterbi ($K = 7$)		
Voice coding	Nil		APC/MLQ
Interleaving duration	640 ms		18.3 ms
Interleaving bits		64.6 bits	
BER voice		10^{-3}	
BER data		10^{-5}	
Es/Eo	2.3		1.0
C/No (no fading) dB/Hz	30.1		44.2
C/No (forward) dB/Hz	35.7		48.6
C/No (return) dB/Hz	35.1		47.0

The two systems, quoted above, relate to what is known as low rate, which can support data only at a bit error rate of 10^{-3} and a high rate which can support both voice and high speed data at a bit error rate of 10^{-5}. A typical configuration is shown in Fig. 6.6 for the overall AES. The voice/data processing is dependent upon the mode of operation; for high rate systems the processing components are as in Fig. 6.7.

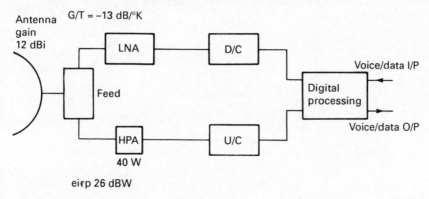

Fig. 6.6 Aeronautical earth station system diagram.

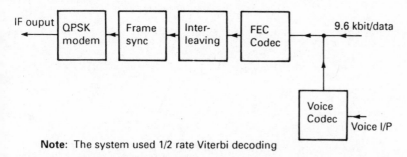

Note: The system used 1/2 rate Viterbi decoding

Fig. 6.7 Aeronautical earth station digital processor.

The voice signal is processed prior to the forward error correction, which is followed by bit interleaving, chosen in this system to be 384 bits i.e. a delay of 18.3 ms for the 9600 bit/sec data stream. The modulation used for the high rate system is not the standard BPSK which allows spectrum regrowth at the output of the high power amplifier but offset BPSK. This is no different in other characteristics such as BER, and is implemented in more complex hardware. It has a greater probability of cyclic slippage at the output of the demodulator and cannot be used with four-phase differential encoding. For the low rate case it is possible to use standard BPSK with differential encoding applied to take account of the increase in phase noise at the lower data rate.

The need to achieve a high bit error rate under very restrictive conditions means that not only is the use of Viterbi encoding necessary but additional coding is necessary to achieve the required performance. In this case an RS code is applied after the symbol interleaving, which is needed to disperse the errors caused by the Viterbi decoder. The improvement in performance afforded by the use of this coding is shown in Fig. 6.8, both with interleaving and without as well as a comparison with Viterbi coding.

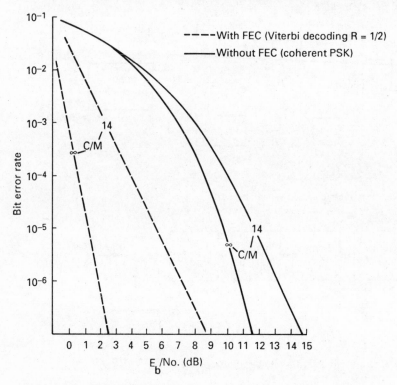

Fig. 6.8 Bit error rate/E/No for Rician fading channel.

A considerable improvement is made if a shortened 3/4 RS code is employed even if no Viterbi processing is utilised, though at the cost of reducing the allowable information rate to 7200 bits/sec. The overall system budget performance, using the INMARSAT 2 satellite, for high and low rate systems is shown in Table 6.8. The forward and return links are both included.

6.3.2 Land mobiles

The mobile satellite services have developed from maritime through aeronautical to land mobiles. The frequency allocations for mobile services on land were set by a **WARC** Conference convened in 1987 and consist of two bands; 1555–1559 MHz and 1656.5–1660.5, with those allocations coming

Table 6.8   INMARSAT aeronautical link budget

Parameter	600 bits		9600 bits	
	Forward	Return	Forward	Return
(a) *Uplink GES/S and AES/S*				
Frequency (MHz)	6.42	1.64	6.42	1.64
Path loss	200.9	189.0	200.9	189.0
Atmos X	0.4	0.4	0.4	0.4
Satellite G/T	− 14.0	− 12.5	− 14.0	− 12.5
eirp	62.0	13.5	62.0	22.5
C/No dB/Hz	75.3	40.2	75.3	22.5
Satellite gain	161.7	158.0	161.3	158.0
Satellite C/No	69.0	59.8	67.0	60.8
(b) *Downlink S/AES and S/GES*				
Frequency	1.54	3.6	1.54	3.6
Path loss	178.8	197.2	188.4	189.9
Atmos X	0.4	0.4	0.4	0.4
AES G/T	− 26.0	30.7	− 13.0	30.7
Sat eirp (dBW)	22.4	− 17.9	22.0	− 8.9
Downlink C/No	36.8	45.1	48.7	54.1
(c) *Link performance*				
Required C/No	33.7	35.7	47.9	47.9
Actual C/No	36.8	38.9	48.6	47.8
Margin	3.1	3.2	0.7	− 0.1

into force in October 1989. INMARSAT will be a provider of land mobile services and their Convention has been amended to include the provision of land mobile services.

The service will not be based on the Standard A terminal, which is too large for most mobile applications even though the system is already in use around the world as truck-mounted equipment such as that developed by Marconi.

The final form of the INMARSAT terminal is still under debate, a debate which includes discussions with other satellite land mobile users such as M-Sat and AUSSAT and deliberations within the CCIR forum to produce a formal definition of a mobile earth station. The initial service will be supported on existing INMARSAT satellites but these will be replaced ultimately by high eirp satellites using spot beam transmissions and therefore allowing the use of smaller earth station terminals. The commercial and technical position of INMARSAT, with respect to land mobiles, is very similar to that pertaining at the start of maritime communications, which used existing satellites, zero customer base and alternative forms of mobile technology such as cellular radio PMR systems and the coming of the cordless telephone. The satellite system again brings to the market place the same advantages as those associated with the maritime systems, i.e. the ability to cover a very wide area without the need to change receiver frequency or earth station configuration. There have been a number of trials carried out on a land mobile system both for normal vehicles and as radio paging services.

Radio paging

A typical system for radio paging is shown in Fig. 6.9. The uplink will be provided by the coast earth station, the information being fed from the radio paging centre, via the PSTN, to the CES.

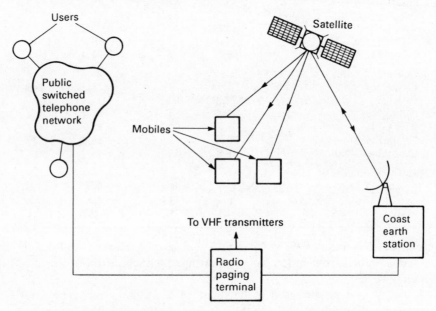

Fig. 6.9 Satellite radio paging system.

System design

The current modulation method used by the aeronautical system is QPSK and it may be that this could be used for a land system. Another modulation method also of use in mobile systems is amplitude modulated single side-band, which has some advantages in that it has no interface problems when linking to the public switched telephone network. The fact that it is a power-limited situation and is operating in a complex fading environment means that one possible form of modulation is multi-level FSK. This is a non-coherent form of modulation and can be realised with simpler circuitry than PSK as no carrier recovery is necessary and the circuitry does not have to have exceptional phase noise performance. If FSK were used then some form of coding must be used to improve the receiver facing performance; an RS code can be utilised as for aeronautical systems.

It should be noted that the fading associated with land mobiles is quite different to that found in other mobile systems; the signal is affected by buildings and moving objects that produce multipath reflections. The 'shadowing' effects by buildings situated between the land mobile and the satellite is the most difficult problem. This effect produces not only amplitude variation but also phase modulation of the received signal which has serious effects on the receiver performance. The effect of shadowing also means that carrier acquisition, if such a system is used, must have the ability to acquire the carrier very quickly and so avoid poor fading performance.

6.4 GLOBAL POSITIONING SYSTEMS (GPS)

The use of satellites to provide a system that will give users their exact position on the earth's surface was developed by the US Navy in the 1960s to support their Polaris submarine fleet. This system, known as Transit, evolved into a more wide-ranging system now known as NAVSTAR, which will provide a very accurate positioning capability, not only for maritime vessels but also commercial vehicles, cars and aircraft, providing a positional fix to within 15 metres. There are three elements to the system:

- Satellite configuration;
- Satellite earth stations;
- System control centre.

6.4.1 Satellite configuration

In order to provide the necessary total earth coverage the satellite system consists of a large number of satellites in non-synchronous orbits. In the GPS this means that 18 satellites will be launched, together with three oribiting spares at an altitude of 20 169 metres; this number should ideally be 24 satellites but for reasons of cost this number may not be reached. The satellites will occupy near circular orbits which are inclined to the equatorial plane at 55°, having a twelve-hour orbital period. The satellites are configured in six separate orbital planes each containing three satellites. This means that from any point on the earth's surface it is possible to see four satellites at any one time, assuming an elevation angle of the earth station's antenna to be greater than 5°.

The number of satellites is critical to the provision of full coverage over the whole of the earth's surface but is a very expensive concept to achieve, in addition, there are some doubts on the system's availability as it is military-based. One alternative to the system described would be to provide a less expensive system, using less satellites, but this is at the expense of the amount of coverage that can be provided; for instance seven satellites will give worldwide coverage for only six hours per day.

The positional information given to the user has a three-dimensional structure, i.e. latitude, longitude and altitude. In order to fix a position the satellites all broadcast a UHF signal containing a ranging code to give the satellite distance from the receiver and positional information to inform the ground of the satellite's position when it transmitted its message. In addition the timing of the signal transmission must be accurately known. To complete the information needed to determine the user's position the ranging process requires not only to know the time of code transmission but also the time of reception on the ground; this is provided by the ranging information of a fourth satellite. The fact that all the satellites are synchronised means that the earth station receiver has to solve four simultaneous equations, each of which has had a clock constant added to it at the input of the receiver.

The signals are transmitted, using spread spectrum techniques, on three frequencies; a primary frequency at 1575.42 MHz and two secondary frequencies of 1227.60 MHz and 1381.05 MHz, though the latter frequency is not for positioning purposes. The civil system utilises 1575.42 MHz and has its own special code which signifies that it is a civil operation, known as the Standard Positioning Service, SPS, with a frequency of 1023 kbits/second. The codes are different for each satellite and are 1023 bits in length, with a 1 kHz repetition rate; this high repetition rate ensures that the ground receiver can easily acquire the satellite signal, so reducing the cost of the receiver. The transmission at 1575.42 MHz contains a second code, which is used for precise positioning and operates at a rate of 10.23 Mbits; this code, known as the precise positioning or P code, is 6 Mbits in length and repeats every 267 days, even though it is in fact reset weekly, with a different position assigned to each satellite. The difference in accuracy between the two signals is shown by the fact that 1 bit of the SPS code is equivalent to 300 m and for the PPS only 30 m.

6.4.2 GPS receiver

Both the C/A and P codes are transmitted at 1575.42 MHz; these codes spread the signals over a 20 MHz band, giving what is known as a spread spectrum transmission, which is very resistant to jamming and can be transmitted at relatively low power. The GPS receiver has to be sensitive to a received signal at a level of -160 dBW, it has to despread the signal by comparing the received signal against a code generated within the receiver and it is the function of the receiver to align this internal code to the received code. This correlation method works by being aware of the amount of the correlation between the codes and so tracking the input signal and by use of the Costas Loop, or its equivalent, it is possible to take account of the Doppler shift due to satellite movement. The error in position measurement will not be uniform over the whole of the earth's surface, with errors being generated by:

(a) Satellite ranging errors, an effect compounded by the use of the minimal number of satellites. For should the satellites that are being seen from the GPS receiver be too far apart then the positional uncertainty increases. The error in the ranging information is due to ionospheric propagation errors which are caused by the refracting of the signal by the ionised gases that exist up to 60 km above the surface of the earth. In effect the refraction alters the signal path length and the signal velocity.

Another atmospheric layer is the troposphere, which extends up to 30 km above the earth's surface. This layer is not ionised but produces refractive effects due to the water vapour in it and this error must be removed from the received signal.

(b) Relativity effects, which are twofold; the first due to the relative velocity between the satellite clock and the earth-based receiver clock – an effect that makes the receiver clock seem to be running slow. The second is due to the earth's gravity affecting the clocks, which can be compensated for by offsetting the satellite's clock frequency.

(c) Receiver errors, due to the errors involved in the demodulation process within the receiver. In addition to this there are the effects of multipath, where the receiver input is both a direct and indirect signal, which has a longer path length and may be at a higher signal strength which will capture the receiver, rather than the true signal. Even if the receiver is not captured the multipath signal may still produce errors due to pseudo-noise effects in the demodulator.

6.4.3 GPS system control

The main function of the control system is to ensure that the satellites are functioning correctly and to update the positional information from each of them. This is accomplished by the use of four remote monitoring stations which collect data from each satellite as they move in their allocated orbit. This information will be transmitted to the control centre which upgrades each satellite with new navigational information (that each earth-based receiver sees) as well as the updating of satellite clocks, timing and signal format information.

6.4.4 GPS development

The role of GPS as the single major worldwide positioning system, used not only for mobile systems but oil production platforms, pipeline laying, cartography etc., is unlikely to be more than a pious hope if the accuracy is not improved and the user capital and rising costs are not kept low. In addition to this it may well be that consumer resistance, such as that experienced in INMARSAT, will delay the widespread use of such a system. However the improvement in accuracy could be achieved by:

(a) The use of a larger number of satellites, though this would increase the system's cost.
(b) New receiving techniques; in essence the options available all have the same purpose i.e. to remove the error due to the system problems. Proposed solutions, such as the differential GPS, turn the system into a feedback system in which the satellite information is also received at a new position which is a fixed point and this information is used to compute the satellite navigation errors and to transmit this information to other mobile users of the system. Further improvement in accuracy could be achieved by interferometry techniques, which use multiple fixed receivers whose positional information data is cross-correlated to check on the received signal phase at each receiver and so determine the distance between receivers to an accuracy measured in centimetres and so ideal for surveying purposes.

6.5   LAND MOBILE SYSTEM (MSAT)

It is necessary to describe the NASA-derived programme for a mobile communications system for the USA, in which the satellite is used to communicate to receivers in motor cars, lorries etc., even though it is not an operational system. It is however the major thrust towards mobile communications in the USA.

6.5.1   MSAT overall system

The MSAT system (Fig. 6.10 shows the basic system) has been under consideration for almost ten years. It began with studies carried out by NASA but has continued with work contracted out to the Jet Propulsion Laboratory (JPL) of California. The aim of such work was to produce a commercial mobile satellite system at low risk both in terms of technological and commercial elements. The proposed system will provide telephone and data services across the whole of the mainland USA and this coverage is known as Continental United States (CONUS). The original design considered the use of the 860 MHz band but due to the new WARC agreements will now be operating in the 1600 MHz or L Band, using transmit frequencies in the range 1646.5–1660.5 MHz and receiving in the range 1540–1559 MHz, with the link between the MNC and base station being set up at Ku Band. To date the design work has concentrated on the earth station characteristics and the current work programme is designated as MSAT-X.

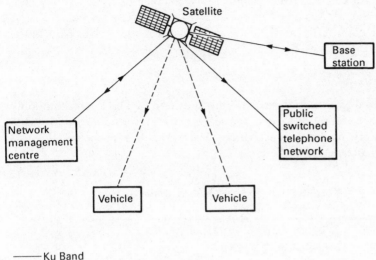

Fig. 6.10 MSAT basic system.

It is envisaged that the initial satellite would employ frequency re-use and two multi-beam antennas, one for transmit and one for receive. To cover the CONUS area will require four beams. The most suitable access method is FDMA and the modulation method will be differential eight-phase PSK with trellis encoded modulation (TCM/D8PSK). This system supports 5 kHz channel bandwidths and is not only spectrally efficient (1 bit/Hz) but provides protection against multipath fading. The first satellite will be leased from an existing organisation.

A typical link budget is shown in Table 6.9, but it is envisaged that a second generation will have antennas three times the size of the initial system

Table 6.9   Typical MSAT link budget

Parameter	Forward link G–S	Forward link S–U	Return link U–S	Return link S–G
Antenna gain dBi	49.6	33.7	7.8	31.2
Transmitter W	1.0	0.8	1.6	0.1
eirp dBW	47.6	31.1	8.5	19.7
Path loss dB	206.9	188.3	188.9	205.8
Rx G/T dB/°K	4.0	− 15.4	9.2	21.0
Rx C/No dB/Hz	66.3	51.9	53.2	56.7
System C/No		51.1		51.1
Required C/No		47.8		47.8
System margin		3.3		3.3

to increase the capacity from 1900 to 12 000 channels by using multiple spot beams. The access system will be demand assigned and use a slotted ALOHA protocol to control the flow of information between the various parts of the system.

Mobile earth stations for MSAT-X

The proposed earth station realisation is shown in Fig. 6.11, which not only takes into account the need to minimise the physical size but also is flexible in concept to provide for system changes. The parameters for the mobile earth stations are detailed in Table 6.10.

Table 6.10   MSAT mobile earth station design parameters

Parameter	Value
Rx coverage	20°–60° in elevation
Channel BW	5 kHz
Rx I/P level variation	< 6 dB rural
	< 8 dB environment
	< 25 dB vegetation
Modulation	TCM/8PSK
Doppler variation	< 1600 Hz
Doppler reduction	± 180 Hz
Operating BER	1 in 10^{-3}
Vocoding	2400 bits
Fading model	Rician (10 dB Rice Factor)
Error control	ARQ punctured convolutional code
Decoding	Variable rate Viterbi
FEC	Rate $\frac{3}{4}$ convolutional
Speech coding	Linear predictive (LPC)

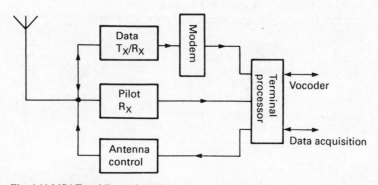

Fig. 6.11 MSAT mobile earth station.

The majority of elements have been reviewed and general results are shown later in the chapter. The Ku Band earth station is standard in form and has the characteristics listed in Table 6.11.

Table 6.11   MSAT hub station parameters

Parameter	Value
Frequency band	12/14 GHz
Antenna diameter	3 m
Antenna gain	50.2/48.8 dBi (Tx/Rx)
Polarisation	Linear
Input noise °K	341 °K
Rx G/T dB/°K	23.3 dBi
HPA power	34 W (base station)
	158 W (PSTN gateway)
eirp	64.3 dB/71 dB

6.6   SATELLITE AIDED SEARCH AND RESCUE (SARSAT)

This system is a unique example of international co-operation between Canada, France, the USA and the USSR. The first system was developed in the early 1980s and was first demonstrated in 1982. It has proved to be a very effective worldwide system which has rescued over 1000 people involved in aviation and marine incidents. The system has five satellites, two SARSAT satellites, launched by the USA and three COSPAS satellites, launched by the USSR. These have a dual role for navigation in the case of the USSR satellites and for meteorology for the USA satellites. The satellites operate at an altitude of 1000 km and are in sun-synchronised polar orbit. In Fig. 6.12 the methodology of the system is shown. This effectively uses the satellites as relay stations which pick up signals from the radio beacons and transmit the message to ground stations which can compute the distress position.

The type of beacon unit used depends upon the application; aircraft are fitted with emergency locator transmitters (ELT), while maritime units are called emergency position indicating Radio Beacons (EPIRB). The beacon unit, when activated either manually or automatically, transmits a signal to a satellite using a frequency in the 121.5 MHz band and also the 406 MHz band. The 406 MHz band is allocated solely for use in the satellite-aided search and rescue, as decided at WARC 7. In addition the frequency will be used for the Global Maritime Distress and Safety System (GMDSS) for maritime applications. The satellites transmit signals down to local user terminals (LUT), which are configured to operate at the two frequencies already quoted and the military frequency of 243 MHz.

The USSR satellites process both 121.5 MHz and 406 MHz signals, while the USA satellites also include the 243 MHz facility. The satellites not only

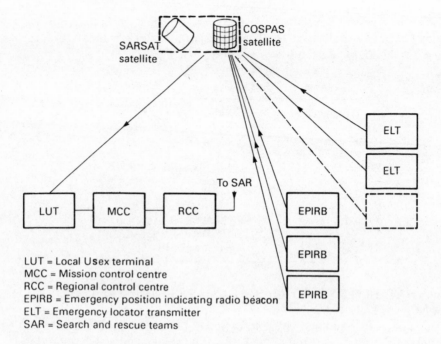

LUT = Local User terminal
MCC = Mission control centre
RCC = Regional control centre
EPIRB = Emergency position indicating radio beacon
ELT = Emergency locator transmitter
SAR = Search and rescue teams

Fig. 6.12 SARSAT system configuration.

act as straightforward repeaters with information at 2.4 kbits in real-time processing in the LUTs, but they also store data on board to allow the retention of 2048 basic messages which are read out and transmitted between real-time transmissions. The mixture of real-time and stored transmissions allows the system to provide total global coverage. The signals from the Soviet satellites are available to anyone who wishes to receive them and Leeds University in the UK has conducted a series of tests using their own receiving system which assumed the link budget detailed in Table 6.12.

Table 6.12   SARSAT link budget

Assumed satellite VHF eirp	+ 10 dBw
Path attenuation for 1000 km	− 136.4 dB
Boltzmann's Constant	− 228.6 dBW/°K/Hz
Ground station G/T	− 25.7 dB
Antenna gain	2.2 dB
Rx noise temperature	120 °K
Antenna noise temperature	520 °K
IF bandwidth	20 kHz
Rx carrier to noise	33.5 dB
Received power	− 124.2 dBW
Receiver sensitivity	− 157.7 dBW

6.7 GLONASS NAVIGATION SYSTEM

The GLONASS system is the equivalent of the US NAVSTAR; it operates in a similar manner to GPS, uses the same type of system parameters and is very close to being operationally compatible to the GPS system. The current operational status is that there are five satellites in use, a total that will ultimately rise to 24 for a fully operational system. The satellites will occupy three orbits, with eight satellites per orbit, separated by 45° of arc along the orbital path. To indicate how close the systems are the comparison is shown in Table 6.13.

Table 6.13   GPS/GLONASS satellite comparison

Parameter	GPS	GLONASS
Tx F1 MHz	1575.42	1597–1716
Tx F2 MHz	1227.6	1240–1260
Coding	CDMA	CDMA
Precision code P Mbits	10.23	5.11
Coarse code C/A Mbits	1.023	0.511
Code repeat time	1 ms	1 ms
Information data rate	100 bps	50 bps

The basic form of the data, transmitted from the satellite, is the same between each system i.e. it must have:

- Ephemeris information for the satellite;
- Timing information to allow system co-ordination;
- Ranging information to give satellite distance;
- General status data.

The difference in the above information is not a general one but a system concept one. In GPS an assumption is made that the orbits are smoothly elliptical, even though this is untrue because of the gravity effects, while the Soviet system represents positive information by ECEF co-ordinates and their derivation in time. The accuracy of the systems is similar though the path to that accuracy is different, for with GLONASS the system update is four times per hour while the GPS system updates twice per hour.

The differences between the systems are not large and collaboration would reduce the cost to each country and would ensure that the system would be fully implemented and therefore provide high accuracy.

6.8 PRODAT MOBILE COMMUNICATION SYSTEM

6.8.1 Introduction

The European Space Agency (ESA) has begun a consideration of a satellite system named PROSAT that would encompass all three forms of mobile communications, i.e. land, sea and maritime. This has a programme of tests and demonstrations relating to mobile communications. The maritime system was titled PROMAR, while the PRODAT system was envisaged as a low cost, small terminal implementation which would provide a data-only service. Its terminals would have a low G/T of -24 dB/°K which could be used for all three applications. The basic design of the terminal therefore must be adaptable to various conditions associated with the communication channel. The system design and equipment specification is now well advanced and trials have been carried out with a system as shown in Fig. 6.13. The overall system uses:

- The MARECS A satellite;
- PRODAT terminals;
- Network management at Villa Franca;
- Fixed terminals.

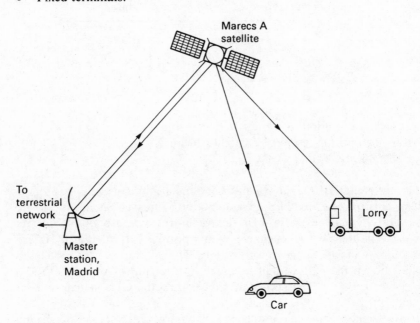

Fig. 6.13 PRODAT experiment.

The system is designed to provide the following:

- Messaging between mobiles and between fixed and mobile;
- Multiple messaging by use of fleet addressing;
- A dialogue mode between users – call termination by either user;
- Polling across mobiles;
- Paging to Rx-only mobiles;
- Network management that will hold a total list of mobiles in the system and their status and also monitor the overall system status, collect and display traffic statistics and undertake billing and equipment monitoring.

The study results led to system parameters as shown in Table 6.14.

Table 6.14   PRODAT system design parameters

| Parameter | Link | |
	Forward	Return
Baud rate	1500	300
Modulation	BPSK/DE	CDMA/DE
Data rate/channel	46 bits	150 bits
Number of channels	15	15
Coding	Short block RS	
Redundancy	ARQ	–
Bandwidth	25 kHz	650 kHz
Frequency accuracy	± 25 kHz	± 600 Hz
Nominal C/No	37–42 dB/Hz	31–36 dB/Hz

The three areas of communication have different environmental constraints, propagation characteristics and speed of transportation. The major effect in aeronautical or maritime channels is multipath and a Rician propagation model is used.

6.8.2   PRODAT mobile terminal

A typical terminal configuration is shown in Fig. 6.14.

Antenna

The antenna coverage differs between aeronautical and land/maritime applications and therefore different solutions must be applied. The aeronautical antenna has to have a gain of 1 dBi and provide 0°–360° in azimuth and 10°–90° in elevation though in trials two antennas are used, one a quadrifiliar helix to cover 30°–90° and a Lindenblad array to cover + 30° to − 30°. The land terminal has to cover + 20° to − 30° and is thus not so difficult to design as an aeronautical; in the trial the helix antenna was

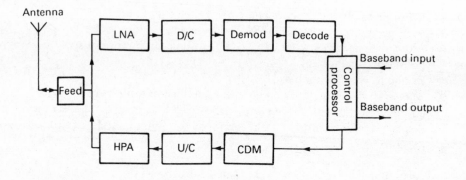

Fig. 6.14 PRODAT earth station.

sufficient. There has been considerable research into this area of mobile communications and steered medium-gain antennas are now considered most appropriate for this application; detailed considerations of mobile antennas are given later in this chapter.

Low-noise amplifier

The low-noise amplifier must provide the required noise figure, high forward gain and low intermodulation, while at the same time ensuring that the physical size of the LNA is small.

A typical specification would be:

Input bandwidth	1530 MHz–1559 Mhz
Noise figure	1.2 dB
Forward gain	50–60 dB
In-band spurious	− 60 dBc
Out-band spurious	− 95 dBm

High power amplifier

The HPA for the PRODAT terminal has to be small in size and could be solid state; a typical specification is given below:

Frequency range	1626 MHz–1661 MHz
Output power	15 W
Input level	+ 5 dBm to + 15 dBm
HPA gain	42 dB

RF sub-system

The RF System must take account of the effects of the transmission medium, such as Doppler, which is much higher for aeronautical systems, as equation 6.1 shows.

$$F_d = 89.4 \left(\frac{V_v}{\lambda}\right) \tag{6.1}$$

where F_d = Doppler frequency shift

V_v = vehicle velocity

λ = wavelength of carrier frequency in cm

For the PRODAT system the Doppler shift seen at the satellite, must be less than ± 100 Hz.

6.9   MOBILE EARTH STATION SYSTEMS

6.9.1   Antennas

When considering the need to reduce both size and cost of ship earth stations the antenna is the major element amenable to improvement. The INMAR-SAT Standard A will retain the parabolic form of antenna, with its complex method of stabilisation etc. but for Standard B and Standard C the development of mobile antennas is towards short backfire, phased array and dipole based antennas. The main forms of suitable antenna will be considered below:

Short backfire antenna

This is very suitable for a medium gain system. The standard form of the antenna is shown in Fig. 6.15, with the dotted additional reflector giving improved gain and axial ratio performance. The two planar reflectors are separated by a half wavelength and these act as a leaky cavity resonator which gives a radiation that is normal to the sub-reflector. The design can have as many elements as 16 and typical performance is shown in Table 6.16.

Quadrifiliar helix antenna

The attraction of this is that it is simple, requires no stabilisation if used in a maritime environment and needs no pointing circuitry. A quadrifiliar helix antenna consists of four helices wound on a cylinder, spaced at an equal distance from one another. The helices are fed with differently phased signals at 0°/90°/180°/270° relative to one another. The antenna is simple to implement and manufacture in that the radiator elements are attached to a flat

A = Main reflector
B = Crossed dipole feed
C = Sub-reflector
D = Improved SPF sub-reflector
a = Half-wavelength distance

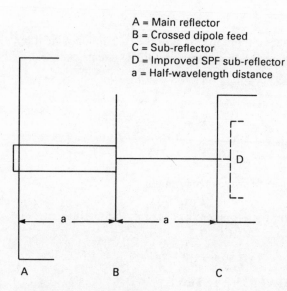

Fig. 6.15 Short backfire antenna.

plastic sheet which is then wrapped round a tube, made up of dielectric material, as shown in Fig. 6.16. The antenna is fed by a microstrip circuit

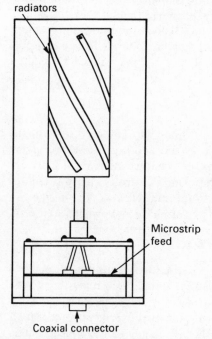

Fig. 6.16 Quadrifiliar helical antenna construction.

which applies the signal to the antenna. Typical performance is shown in Table 6.15 and this can be characterised as a medium-gain antenna.

Phased array antenna

This is considered to be a medium-gain antenna and can be implemented in a flat form. A typical antenna, produced by Teledyne, is made up of 19 basic elements, the construction of which is shown in Fig. 6.17. The element consists of:

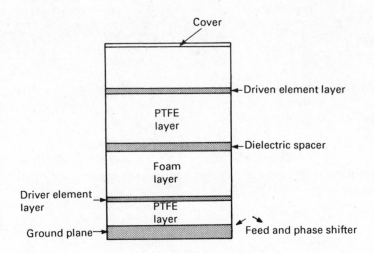

Fig. 6.17 Phased array antenna construction.

- Phase shifters;
- Microstrip feed;
- Driver element;
- Dielectric spacer;
- Driven element;
- Covering.

The overall antenna is comprised of 19 of the above elements. The microstrip feed network provides the power dividers which connect to the 45°/90°/180° phase shifter elements. The phase shifter output is connected to the driver elements, which are used to give circular polarisation. The driver element is separated from the driven element by a PTFE dielectric spacer, whose thickness and characteristics determine the VSWR and pattern form.

The great advantage of the phased array is that the antenna can scan the beam pattern and therefore can be used as a tracking system. A typical antenna will scan every position at 15° in azimuth and two elevation points. The scanning takes place every two seconds, the acquisition period being set by the need to minimise multipath and fading effects. The other form of this antenna is simpler in that it uses only four patches connected to form a linear array, which is tilted at 45° to give a CONUS coverage as required. The tracking in this case is achieved by connecting the patches to give sum and differences as in monopulse tracking system. Table 6.15 shows a typical performance.

Table 6.15   Mobile antenna performance comparison

Ae Type	Gain (dBic)	°K	Hgt/Dia	S/IdB
Crossed dipole	4	50	5/3	10
SPA	8	205	1/22	25
Tilt array	12	190	9/20	24
SBF	15	–	3/15	–

Dipole antenna

This is usually in the form of a crossed dipole which is made up of two dipoles crossed orthogonally at their centres. The radiation at low elevation angles of such a configuration can be increased by allowing the individual dipoles to droop down.

6.9.2   Diversity requirements

The use of smaller antennas for all forms of mobile system has meant that the fading of the transmission medium has assumed a much greater significance and a variety of methods can be employed to counteract this effect, which can give as much as 20 dB variation in signal level to the receive system.

The normal means of obtaining diversity, frequency, space, polarisation or angle are not all relevant to the mobile environment. For ship earth stations the possible methods are as follows:

Polarisation

Polarisation diversity can be applied because of the peculiar nature of the signal reflections from the sea, where the horizontal polarisation remains unchanged regardless of sea conditions or elevation pointing angle. The phase angle of the vertically polarised wave meanwhile, continuously changes as the elevation angle increases. In addition, above 6° elevation, the circularly polarised wave becomes elliptical after reflection as well as reversing its sense of rotation. Reflected wave suppression can be achieved by ensuring that the antenna polarisation ellipse remains orthogonal. This system can be implemented using an SBF antenna that uses variable phase shifter circuits to control the antenna characteristics. This form of diversity can achieve fading depth reductions of up to 3 dB.

Maximal level tracking

This assumes that it is possible to control the antenna radiation characteristic in the direction of the reflected wave. If this can be done then a fading depth reduction of 5 dB can be achieved. In practice a quadrifiliar helix can be used where variable phase shifters are inserted into the feed circuit of each element. The control circuitry senses the received level and changes in that level when slight adjustments are made in the elements' input phase. If the level increases then the phase is changed even further in the same direction, if not then the phase is reversed to move it in the opposite direction.

Pattern shaping diversity

Pattern shaping diversity can give considerable fading reduction but also gives a loss of antenna efficiency. This method shapes the main beam to be a flat form, while suppressing radiation in all other directions.

6.10 SIGNAL PROCESSING

The trend in mobile satellite communications is the same as that in fixed services, i.e. digital operation. There is therefore a need for the use of signal processing to achieve the required performance in terms of data rate and performance quality. To this end the signal is coded and digital modulation techniques are used.

Coding

The coding used by PRODAT is a short RS code which encodes as shown in Fig. 6.18.

The block code is realised in both horizontal and vertical dimensions. The horizontal dimension forms one block of n vectors, where the primary vectors consist of k' symbols and redundancy vectors consist of $n - k'$ symbols, which correct burst errors, due to the inherent interleaving characteristics of the code. In the vertical dimension one RS code forms a vector made up of k symbols of data and $n - k$ symbols for redundancy purposes, to correct random errors and detect bursts of errors that the code cannot handle. The RS code can be combined with a classical ARQ scheme to improve the performance when the vertical vector has been erased. The delay of the satellite link means that the code blocks have to be numbered and the return channel used with care as it is no better in performance than the forward channel.

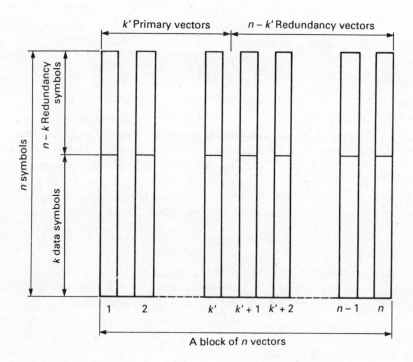

Fig. 6.18 PRODAT RS code.

The basic principle of the scheme, patented by ESA, is as follows: the time axis has a Rate 1/2 code (16:8), giving k' of eight vectors, which means that the code can correct up to eight errors. The loss of a vertical code vector can be due to either a deep fade or because the system error correction ability has been exceeded. The decoding process decides if a vector is correct or has been erased; with a 16:8 code it is possible to decode a vector as soon as eight vectors have been correctly received. The foregoing explanation applies to the forward link, for the return link the system is modified to send eight redundancy vectors as a single block if any errors have been received in the previous block. Acknowledgement is made during the transmission of the next block and should this not be the case the transmission may be re-transmitted or interrupted.

The RS code is often used as an outer code with a convolutional code to give improved performance, though this will require more complex codec equipment.

The modulation method

This method used on the Standard A INMARSAT system, was narrow-band frequency modulation which had the characteristics detailed in Table 6.16.

Table 6.16   Narrow band FM parameters

Parameter	Value
Baseband frequency	300–3000 Hz
Pre-emphasis	Not applied
Companding	2:1 syllabic (CCITT G162)
Peak frequency deviation	12 kHz at 0 dBmo BB level
Mean speech level	− 14 dBmo
Standard deviation	6 dB

The cheapest and simplest method of providing HPA power is to use a Class C amplifier and therefore it is important that the method of modulation produces a constant amplitude signal in order to avoid high power sidelobes, produced by the non-linearity of the amplifier. The bandwidth of the system is limited and therefore the modulation must also be spectrally efficient in order to conserve bandwidth. The most used modulation scheme is phase shift keying at various levels. Standard QPSK is unsuitable as it does not have a constant amplitude characteristic. Offset QPSK however, gives an improvement and a further improvement can be made by using FFSK, i.e. fast frequency shift keying (FSK with a deviation ratio of 0.5). The BER performance is shown in Fig. 6.19.

6.11   TRANSPORTABLE TERMINALS

6.11.1   Introduction

The previous systems covered have been concerned with special systems set up specifically to provide mobile services. There is however one form of mobile application that has used existing satellites for the purpose of news gathering from remote locations. The first terminals were put into operation in the USA in the early 1970s using C Band transponders, 6-metre diameter parabolic antennas mounted on trailers along with all the necessary electronics equipment. These systems required trailers that were 42 feet in length and represented a very large package and considerable capital investment.

The advent of Ku Band satellites and an improvement in antenna design has led to considerable reduction in the size and weight of transportable terminals, both in the van type and the so-called flyaway systems which allow the earth station operator to carry the total terminal on a normal commercial airline. The overall system is shown in Fig. 6.20.

The system diagram shows the four elements of any system; the satellite which must be visible from the earth station positions and have the correct frequency of operation; the satellite operations centre or hub station which is used to set up the television link and control the transmissions from the

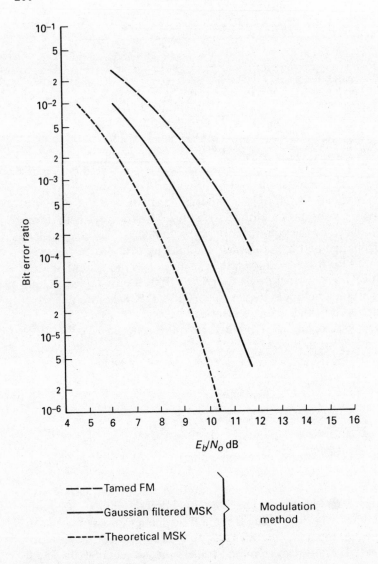

Fig. 6.19 Bit error rate/signal ratio.

portable terminal; the portable terminal itself and the remote receive-only terminal, situated at the television studio, which is receiving the television pictures from the remote site. The typical portable terminal has a form shown in Fig. 6.21.

The basic characteristics of the flyaway terminal must take account of the following requirements:

(a) The terminal must be as light as possible and made up of packages that

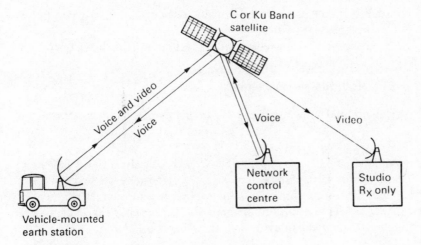

Fig. 6.20 Typical satellite system using transportables.

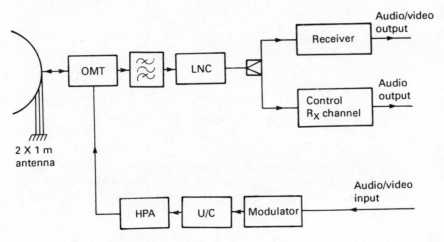

Fig. 6.21 Portable terminal configuration.

are not only low enough in weight to allow it to be carried but are also of manageable proportions. This requirement is not only to ensure that it can be carried but also to reduce operating costs by reducing the transportation charges.

(b) The terminal must be simple to operate not only because in general the level of technical expertise is lower but also because the setting up of the terminal can be very expensive if it takes any length of time.

(c) The equipment has to be able not only to withstand a very wide range of environmental conditions but also it must be mechanically very strong in order to withstand the severe handling conditions that it will encounter, such as travelling over rough terrain or being dropped from vehicles and aeroplanes.

(d) As the equipment is non-redundant it must be very reliable in operation.

6.11.2 A typical transportable terminal

The typical terminal shown in Fig. 6.21 is made up of a number of packages.

Antennas

The antenna must meet the appropriate INTELSAT, EUTELSAT or FCC specifications. The form of the antenna is likely to be an offset feed system, in order to give maximum efficiency, with a reflector that is either elliptical or square, which will give maximum gain for the given surface area. The performance of a typical antenna is given in Table 6.17.

Table 6.17   Typical parameters of transportable antenna

Parameter	Value	
Tx frequency GHz	14.0–14.5	
Rx frequency GHz	10.95–12.7	
Sidelobe pattern	$29-25 \log \theta$	$1° < \theta < 36°$
	-10dB	$\theta < 36°$
Tx gain	45 dB	
Rx gain	43.8 dB	
Polarisation purity	-35 dB	
G/T	19 dB/°K	

The most important parameters are the sidelobe performance and the polarisation purity. The sidelobe performance is the determining factor with regard to the maximum eirp and therefore sets the ability of the terminal to communicate from any point in the world as well as the amount of traffic that can be supported. The LNA is often connected as part of the antenna package, though the frequency of the unit will differ, depending upon the area of use.

High power amplifier

The high power amplifier used in the flyaway system is dependent upon the satellite in use; typical link budgets are shown in Table 6.18. The actual limit on the amplifier is as much a matter of weight and physical size in this application and leads the designer not only to equipment that is small but is lightweight as well. This particular problem is often overcome by splitting the amplifier into power supply and amplifier sections. The other factor that allows the user to restrict the size of the output power is that the required availability is not as high as that needed for normal commercial operations. Obviously the major parameter is the eirp, rather than HPA power, and therefore the antenna size becomes of crucial importance. The other element in considering the trade-off in performance is if the system will transmit more than one TV carrier. If this is not the case then the HPA can be operated closer to saturation and thus allow the use of a smaller basic unit. A further element to consider is the amount of primary power required which will have to be provided by some form of portable generator.

Table 6.18   Typical link budgets for transportable system

Parameter	Domsat		ECS
Antenna size (m)	3.7	3.7	2
E/S eirp dBW	69.0	69.0	69.0
Propagation loss (dB)	207.6	207.6	207.6
Satellite G/T (dB/°K)	0	0	− 1.0
Uplink C/T (dBW/°K)	− 138.6	− 138.6	− 139.6
Sat eirp (dBW)	47	47	40.8
Downlink loss	205.7	205.7	205.7
E/S G/T (dB/°K)	29.9	19.0	39.0
Downlink C/T (dBW/°K)	− 131.8	− 137.7	− 133.1
Overall C/T (dBW/°K)	− 139.4	− 142.2	− 140.5
Boltzmann's Constant (dBW/°K/Hz)	− 228.6	− 228.6	− 228.6
C/No (dB/Hz)	89.2	86.4	88.1
Rx bandwidth (dB)	74.8	74.8	74.8
C/N (dB)	14.4	11.6	13.3

Electronics package

The flyaway system is mostly used in a network situation and therefore the baseband system must be able to modulate the audio control signals onto a 6.5 MHz carrier before being combined with the video signal. This signal is pre-emphasised and has added to it the energy dispersal signal before being modulated onto a 70 MHz carrier which is bandwidth limited and then upconverted to the 14 GHz transmit band. The baseband system has the characteristics set out in Table 6.19.

Table 6.19   Transportable baseband parameters

Parameter	Value
I/P frequency	25 Hz–8.2 MHz
Video I/P level	1V p–p CCIR Rec 405–1
Deviation	16–25 MHz/V
Audio sub-carrier F	6.2 MHz and 6.8 MHz
Audio I/P level	775 mV
I/F bandwidth	36 MHz
O/P power	− 12 dBW

The electronic packages can all be carried as separate items, producing an overall system that is made up of up to 15 units.

Chapter 7

Satellite Business Systems

7.1 INTRODUCTION

The majority of satellite communications traffic originally consisted of voice communications and even today, though the method of transmission may change, the largest use of the system is for voice. However the increasing digitalisation of telecommunications, worldwide, and the continuous process of finding more efficient methods of utilising satellite technology has led to greater use by the business community of satellite links.

Constraints on the use of satellite communications have included the lack of direct access to the satellite itself. Apart from the USA, which has private satellite systems, the only route of access to the use of satellite communications has been either the signatory to each satellite organisation, whether INTELSAT, EUTELSAT, INMARSAT etc. or if a domestic system is available, such as INSAT, PALAPA, BRAZILSAT etc., via the satellite operator. The satellite signatory is usually the PTT of the country, operating not only the external but also the domestic communications, and in many ways it is entirely natural, given the nature of satellite communications, that control is exercised by a single authority. This ensures that the satellite is neither damaged by excessive power transmissions nor blocked by earth station operator misuse and that terrestrial communications and other satellite systems do not suffer from interference. These considerations meant that any potential user of satellite communications could not access any satellite, even if his communication requirements were not being properly met by the official satellite access point.

The changing scene in the regulatory process has now altered the opportunities available to the prospective satellite user who sees satellite communications as the most technically and commercially viable means of communication for his particular requirements. The process of deregulation, begun in the USA, has spread via Great Britain to Europe and the Far East, allowing not only far more facilities to be available, in competition with terrestrial services, but also in the setting up of competition in the provision of those satellite services.

7.2 BUSINESS REQUIREMENTS

The business user has a variety of service needs, including voice communications, and data transmissions.

7.2.1 Voice communications

Voice is the major element in any communications link which can be carried digitally, using a 64 kbits data circuit. The quality of the speech circuit depends upon the bit rate and for high quality links the full 64 kbits would be used. There are a number of methods for circuit multiplication that will lower link quality but only require 32/16 kbits.

7.2.2 Data transmission

The need to link from business to business and site to site has produced a wide variety of methods, many of which are data-based and can use any digital medium for the transmission of the information. The uses of data links are:

Teleprinters

Developments in newspaper production gave rise to the use of wordprocessing facilities by journalists allowing electronic production of the newspaper from the original copy through editing to the final input for newspaper printing. This has meant that not only does the need for 'hot metal' typesetting disappear but the newspaper can be distributed to remote printing centres, thus simplifying the physical distribution of the hard copy.

FAX

Here the distribution is done using facsimile machines that operate at 9.6 kbits. This type of operation is being extensively put into operation with the Apollo system set up by EUTELSAT in Europe.

7.2.3 Basic business satellite system requirements

The major difference in the implementation of a business satellite earth terminal and the standard systems used for global communication is that the earth station is situated on or near the user's premises and data is the element used for communication, rather than voice (though this is not excluded).

The overall system thus comprises:

- A satellite;
- A system management centre;
- A user's earth station;
- A satellite terrestrial interface;
- A terrestrial connection or tail;
- Customer premises equipment.

The satellite characteristics will determine the earth station design and particular systems will be considered later in the chapter and in chapter 8 on VSATs. The earth station itself must be equipped with the appropriate interfaces that will allow connection to the various customer equipments. Typical interfaces would be:

(a) A 2048 kbits interface where the data for the $n \times 64$ kbits channel are carried in the first n time slots after TS 0 (the remaining time slots being set to a logical 1);
(b) A CCIT X21 interface for $n > 1$, used for short terrestrial interconnections;
(c) A 2048 kbits interface suitable for videoconferencing. In addition if the system is used for a T1 network then 1544 kbits and 56 kbits must also be catered for.

7.3 APOLLO DISTRIBUTION SYSTEM

The origins of the APOLLO system reside in the development of EURONET, which was designed as a European Community-wide interconnection of bibliographical and other databases throughout Europe. The APOLLO system utilises the EUTELSAT satellite and a system diagram is shown in Fig. 7.1.

The transmission system operates in a store forward mode, where the data is stored in the store data station awaiting satellite capacity. The data sent from the satellite is held in the sink data stations and the control of the information flow is accomplished by the Satellite Access Controllers (SAC) for the transmit system and by the Receive Access Controllers (RAC) on the receive side. The data stations pass the information requests, commands etc. by using the terrestrial network.

The delivery system design must have the following characteristics:

- Up to 2048 kbits data rates;
- Very low error rates, essentially error-free;
- Low cost of transmission;
- Point to multi-point transmission capability;
- Low cost of capital investment;
- Flexibility to allow for growth.

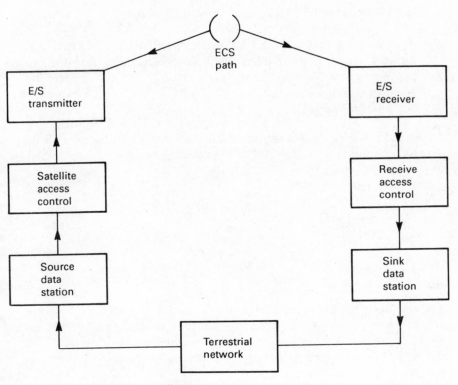

Fig. 7.1 APOLLO system diagram.

7.3.1 Basic design parameters

In the APOLLO system users have only a one-way system, involving them in installing receive-only equipment packages. A typical earth station system is shown in Fig. 7.2.

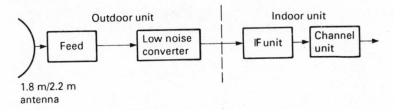

Fig. 7.2 APOLLO receive only terminal.

The earth station comprises two elements:

(a) An outdoor unit consisting of a 1.8 m or 2.2 m parabolic antenna and low noise converter block, that translates the signal which is in the 12.5–12.75 GHz range.

(b) An indoor unit that recovers the signal transmitted from the satellite. In carrying out this function it has to be able to select the carrier, demodulate the signal, decode the forward error correction, recover the synchronisation to the Satellite Multiservice (SMS) multi-frame and descramble the signal.

The basic specification is set out in an ESA document detailing all the necessary parameters to allow overall system design.

Figure of merit

Antenna size	Condition dB/°K		
	1	2	3
1.8 m	20.6	20.0	19.6
2.2 m	22.4	21.9	21.4

Antenna

The antenna will be a simple fixed mount parabolic reflector, but will allow azimuth and elevation adjustment of 150° for azimuth and 10–50° for elevation. In addition the system must have a fine adjustment mechanism that will allow the operator to align the antenna to the satellite to within 0.4 dB of maximum level.

The off-axis gain is defined as:

$$D = 13.5 + 25 \log \phi \quad \text{for 1.8 m antenna} \tag{7.1}$$
$$= 15.3 + 25 \log \phi \quad \text{for 2.2 m antenna} \tag{7.2}$$

The above applies for off-axis angles between 1° to 20°, beyond 20° the discrimination should be 46 dB for the 1.8 m and 47.8 dB for the 2.2 m antennas. The antenna must be able to receive the signal on a linearly polarised transmission with a polarisation discrimination of at least 30 dB and the ability to adjust the polarisation to maximise the signal received.

Power flux density

The PFD is defined as − 137.5 dBW/m².

Gain/phase characteristics

- Gain/frequency response must be held within 0.4 dB
- Gain linearity defined as + 15 dB for LNC and 0 dBm for the I/F unit for the third order O/P.
- Group delay must remain within 10 ns over any 2 MHz portion of the frequency band.

● Phase noise is set as in Table 7.1:

Table 7.1 APOLLO phase noise specification

Off carrier frequency	Phase noise
1	− 30
10	− 65
100	− 90
1000	− 100

Indoor unit

The input frequency will be in the 1200–1450 MHz band at a level of − 41 dBm to − 54 dBm. Each carrier is allocated on a 22.5 kHz grid. In practice this is achieved by use of a frequency synthesiser adjustable in 200 kHz steps over the input bandwidth of 250 MHz. The carrier frequency variation can be up to 300 kHz and this must be taken into account by the use of a search facility which will sweep the local oscillator over a ± 360 kHz range.

7.3.2 APOLLO receive system

Figure of merit

The system definition puts forward three different conditions of sky noise temperature and the system must be designed to operate under those three conditions. The antenna noise temperature is defined in equation 7.3.

$$T = (1 - a)T_b + 150a \quad °K \tag{7.3}$$

where a = amount of received power in the sidelobes

T_b = sky noise temperature in direction of the satellite

The antenna noise temperature can then be calculated from equation 7.4.

$$T_t = \frac{[T + t(A_f - 1)]}{A_f} \quad °K \tag{7.4}$$

where t = actual feed temperature °K

A_f = attenuation of the feed

The details of the figure of merit are shown in Table 7.2

Table 7.2   APOLLO E/S antenna noise temperature

Sky noise (T_b)	T_r (1.8 m)	T_r (2.2 m)
20	53	48
70	92	89
115	126	125

The overall receiver noise temperature is calculated from equation 7.5.

$$T_r = 290(F_1 - 1) + \frac{t(L - 1)}{G} \quad °K$$

$$+ 290(F_2 - 1) \times \frac{L}{G} \tag{7.5}$$

where F_1 = LNC noise figure, assumed to be 2.5 dB

G = LNC gain, assumed to be 50 dB

L = cross site loss, for 100 m this is 23 dB at 1450 MHz

F_2 = IF unit noise figure, assumed to be 15 dB

t = ambient temperature

For this system the overall receiver noise temperature is 244 °K. The overall figure of merit is shown in Table 7.3.

Table 7.3   APOLLO figure of merit for specified antennas

Parameter	1	Condition 2	3
1.8 m antenna gain dBi		45.3	
2.2 m antenna gain dBi		47.1	
1.8 m system noise temp °K	297	336	370
2.2 m system noise temp °K	292	333	370
G/T 1.8 m antenna	20.6	20.0	19.6
G/T 2.2 m antenna	22.4	21.9	21.4

Indoor unit parameters

A typical indoor unit block diagram is shown in Fig. 7.3.

The first IF frequency is applied to the input of the indoor unit and has first to be filtered to give protection against interference from carriers that may be 6 dB higher in level than the normal APOLLO carrier. The signal is then downconverted to 70 MHz, while limiting the signal bandwidth to

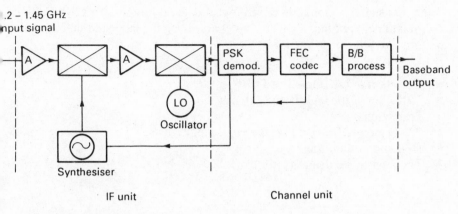

Fig. 7.3 Indoor unit detail.

18 MHz. The APOLLO system operates in a burst mode in order to be compatible with the existing SMS system. It is not however, exactly the same as the SMS system in that it is not a true continuous mode but only a quasi-continuous mode operating with long transmission bursts of up to 10 Mbit/s, with additional bits for a synchronisation preamble, which is 65 536 channel symbols long, corresponding to two SMS multiframes. The preamble allows the receiver to lock on to the SMS multiframe structure.

The PSK demodulator has to provide a number of different functions as listed below:

- It has to recover the carrier signal;
- It must produce the frequency sweep to overcome the satellite frequency drift, to allow carrier acquisition;
- It must incorporate the AGC system that will keep the input to the demodulator at a constant level;
- It must recover the clock signal;
- It must translate the input frequency down to baseband.

The crucial performance parameter that must be met is signal acquisition, which must be met in order to prevent errors. In this case initial acquisition is within 30 seconds and 19 ms from burst to burst.

The importance of error-free operation necessitates the use of forward error correction, which when applied to the transmitted signal gives a channel bit rate that at 3276.8 kbit/s is twice the data rate. The FEC decoder has first to synchronise with the demodulator output before decoding the signal.

The baseband processor performs a number of functions such as descrambling the data signal, a process that is complicated by the fact that the composite signal is a mixture of scrambled data and framing, together with non-scrambled framing. Apart from descrambling the baseband unit has to perform the following functions:

- Detection of the frame word pattern;
- Detection of the multiframe pattern;
- Frame timing;
- Production of frame and decoding timing signals;
- Data buffering of the burst signal;
- Tracking of the demodulator clock between bursts.

7.4 INTELSAT SYSTEMS

There have been a number of developments in the INTELSAT services to take account of the new trends and requirements. The corporate communication network service known as the INTELSAT Business Services (IBS) provides digital carriers operating in an FDMA environment and using QPSK modulation. It is totally dedicated to data communications and excludes telephony. The system provides two grades of services differentiated by the power margin available, which is achieved by increasing the eirp.

The type of earth station used in the IBS system depends upon the country of use, due to the regulatory environment. In a PTT situation, where the access to the satellite is through a single authority, then that authority will only allow the use of national gateways which connect to the customer's premises by the terrestrial network. In environments with a lighter regulation, such as in the UK or the USA, then smaller earth stations can be used. In an unregulated environment the earth stations may well be actually situated on the user's premises, with the consequent need for small earth stations.

INTELSAT has specified two basic forms of earth station, Standard E and Standard F, with these being further sub-divided into three individual specifications to allow the user to select the best type of earth station for their own specific application. The specification is set for both C Band and Ku Band and allows for data rates from 64 kbits to 8448 kbits. In addition to the basic parameters of each earth station there is an allowance made for each user's system requirements, where it is possible to operate either in a closed or open network. An open network is one in which there is a general access to the IBS service. However the earth station parameters and general operating conditions have to be defined to ensure interoperability. In the closed network the user is not subject to restriction with regard to interconnection between earth stations but is only restricted, as it were, internally by setting the RF transmission parameters. In fact the only external regulation applied to the closed network is concerned with the prevention of interference between systems.

7.4.1 IBS transmission characteristics

The IBS earth stations are defined firstly by the figure of merit, G/T dB/°K and cater for both C Band and Ku Band.

Ku Band earth station

INTELSAT Standard E	Figure of merit
E1	$25.0 + 20 \log f/11$
E2	$29.0 + 20 \log f/11$
E3	$34.0 + 20 \log f/11$

In addition it is necessary to know the Standard C figure of merit which is:

$$G = 37 + 20 \log \frac{f}{11.2} \text{ dB/°K}$$

C Band earth station

INTELSAT Standard F	Figure of merit
F1	$22.7 + 20 \log f/4$
F2	$27.0 + 20 \log f/4$
F3	$29.0 + 20 \log f/4$

The Standard F earth stations communicate to Standard A and Standard B earth stations and the characteristics are shown in chapter 5. The above figures refer to those services provided for basic and super IBS and IDR services when there are clear sky conditions. The parameters are also defined to take account of downlink degradation due to rain and an increase in receive system noise. In this case the factor x has to be added to the figures for the Ku and C Band earth stations given above where x is defined as:

$$x = D_{dl} - D_{dv} \tag{7.6}$$

where D_{dl} = local downlink degradation due to local rain

D_{dv} = degradation shown in Table 7.4.

Received power flux density

The RPFD is defined by INTELSAT for the worst case conditions and for each satellite and polarisation, with the specification covering the complete

Table 7.4　Standard F degradation factors

Satellite orbital location	Spot beam	D_{dv}	% yr margin
325.5–341.5	W	13	0.03
174–180	E	11	0.01
307.6–310	W	13	0.02
	E	11	0.02

receive band for Standard E and Standard F. The RFPD for each satellite and spot beam is given in Table 7.5 for both standards.

Table 7.5　Standard E/F received power flux density

Satellite	Power flux density dBW/m^2			
	Standard E		Standard F	
	11 GHz Spot	12 GHz Spot	A Polarisation	B Polarisation
5	− 116.3	–	− 124.3	–
5A	− 112.3	–	− 130.2	− 132.0
5A (IBS)	− 112.3	− 116.8	− 122.4	− 122.4
6	− 110.4	–	− 123.5	− 124.0

7.4.2　Standard E and Standard F earth station characteristics

The INTELSAT Standard E and Standard F earth stations provide the user with a wide range of facilities and can be implemented in a variety of ways.

The antenna sizes can range from 3.5 m to 13 m, with a wide range of antenna/LNA combinations being possible in order to achieve the required G/T. The specification of the earth station and the applications that they are being directed to means that the business terminal is ideally suited to the application of a building block concept for the earth station, in its overall design and construction. This allows the earth station designer to divide the system into three main communication packages:

- Antenna/LNA/ground communication equipment (GCE);
- High power amplifier (HPA) package;
- Baseband system.

The idea of a building block method for designing and building earth stations is not only ideal for the initial installation but can also be used to ensure that the earth station can grow, as the overall traffic capacity grows, without the system having to be redesigned.

The required earth station parameters are calculated in order to achieve a transmission capacity of $n \times 64$ kbits where n can be between 1 and 30. It is possible to transmit a higher capacity if the antenna size and HPA output are increased. The great advantage of this system is that the initial capital costs can be kept to a minimum to cater for the immediate requirement while still retaining the ability to expand the system at a later date.

Any terminal design is a trade-off between performance and cost, though in practical designs a designer often takes a standard antenna that is part of a manufacturer's product range and matches the LNA and HPA to that antenna characteristic. In the INTELSAT specification for SCPC/QPSK systems the antenna system is set out as in Table 7.6.

Table 7.6 INTELSAT E/F earth stations typical antenna gain

Earth satellite standard	Antenna gain dBi
E1	52.0
E2	55.5
E3	59.2
F1	47.7
F2	51.6
F3	53.2

Table 7.6 shows the antenna size, specified by INTELSAT for the IBS service; it should be noticed that the antenna gain used for the figure of merit calculation would be not the same as the actual antenna gain, as it must take into account the losses between antenna and LNA.

The equivalent isotropic radiated power (eirp) requirements are set out in full in the IESS 309 document. The eirp is defined for a range of information bit rates, for Standard E and F earth stations, when using the following satellites:

INTELSAT 5 F1–F4

INTELSAT 5 F5–F9 and F10–F12

INTELSAT 5A F13–F15

INTELSAT 6 F1–F5

The eirp is defined for earth stations that are situated at the edge of the satellite footprint and assumes that the antenna is operating at an elevation angle of 10°. There are also a number of correction factors given which allow the system designer to take account of the possible beam connections and the physical position of the earth stations within the footprint. It is important for the designer to know not only the earth station position but the satellite mode of operation, which beam is in use and the form of interconnectivity, as well as the satellite bandwidth that is being used.

The eirp stability has to take into account two conditions:

Clear sky

Here the variations can be:

Earth Station Standard	Delta eirp
A, B, C, F3	± 0.5 dB
E1/2/3	± 1.5 dB
F1/2/3	± 1.5 dB

The above eirp change is made up of the various factors added on an rms basis.

Adverse environmental conditions

For the C Band system the eirp may fall by a further 2 dB from nominal. For Ku Band the fall in eirp is defined in IESS 309 and ranges between 5 dB and 7 dB. This deviation can be overcome by some form of up-path power control which then has to hold the eirp to within ± 1 dB, with a response time of 1–2 seconds.

Earth station digital parameters
The earth station terminal has to be defined not only in RF characteristics but also in terms of the digital parameters, these are:

(a) Modulation method must be quadrature phase shift keying (QPSK).
(b) It is necessary to ensure that the carrier spectrum is spread at all times during transmissions. In IBS this is done by scrambling as CCITT Recommendation V35, or by any system that can limit the power flux density on the earth's surface to CCIR Recommendations 358-3. This requirement is equivalent to the energy dispersal that is applied to the transmitted signal for analogue systems. The scrambling is applied to the signal as mentioned in the APOLLO system description in section 7.3.
(c) Forward error correction in the form of rate $\frac{1}{2}$ convolutional encoding with Viterbi decoding must be applied to the signal. The FEC is necessary to allow the system to meet the bit error rate performance which is defined as shown in Table 7.7. The Eb/No applies to the modulated carrier power and the data rate is equal to the information rate plus an overhead.
(d) Encryption may be applied to the signal if the user wishes to ensure that his system is secure. This is not mandatory under the satellite system definition but it must not have a deleterious effect upon the transmitted data.

Table 7.7 **Standard E/F bit error rate**
requirement

BER	Information bit Eb/No dB
10^{-3}	4.2
10^{-4}	4.7
10^{-6}	6.1
10^{-8}	7.2

(e) The carrier frequency variation is important in the digital system in that it determines the acquisition design of the baseband demodulator and must be kept to $\pm$ 0.025 $\times$ rate Hz, though it can go out to $\pm$ 10 kHz. The other area of frequency error is in the satellite due to the transponder translation process and this can be $\pm$ 25 kHz or $\pm$ 42 kHz depending upon the satellite in use. There is one parameter that is most important in digital transmission systems and that is the transmission delay or to be more specific, the variation in that delay due to satellite motion. This is actually defined in IESS 309 to allow the receiver to include enough storage to overcome this change of delay.

The general earth station must encompass:

- The transmission rates to be used;
- The use of quadrature phase shift keyed modulation;
- The use of forward error correction techniques, with $\frac{1}{2}$ or $\frac{3}{4}$ rate coding;
- The eirp necessary to carry the traffic density required, together with the off-axis performance specified by INTELSAT.

7.4.3 Intermediate data rate service

Introduction

The Standard B earth station was originally designed for telephony but with the increasing need for digital communications the Standard B specification was modified to allow a data carrier system that could provide the earth station user with reasonable data rates, though not as high as those associated with TDMA. This service was known as Intermediate Data Rate (IDR). The INTELSAT organisation has been developing a 2048 kbit service since 1979 and during that time the specification has been steadily evolving towards an ever more flexible system. Originally the IDR service was available to those users communicating between Standard A, B and C earth stations. Error correction was only optional and the engineering service circuits (ESC) were not completely defined. By 1986, due to user pressure, the service was expanded to include Standard E and Standard F, i.e. the smaller earth stations, as well as bringing it into line with the new developments in ISDN; the use of the smaller earth stations made the use of forward error correction a necessity and in IESS 308 the user was allowed to use $\frac{3}{4}$ rate FEC.

The success of the IDR system meant that the specification has had to be defined for bit rates up to 45 Mbits and the ESC system requirements set out to allow its use on digital systems up to the 45 Mbit data rate; the effect of these changes has been to increase the data rate transmitted and INTELSAT have defined the necessary overhead structure.

The latest issue of IESS 308 is Revision 3 and it provides the vehicle by which the satellite service can access the PSTN. It is compatible with ISDN performance. The competition for IDR in satellite terms is TDMA and in order to ensure that the service is cost-effective it also uses the same technique as TDMA to reduce channel costs, i.e. digital circuit multiplication (DCME). The use of DCME increases the channel capacity by a factor of at least four. An important element in specifying the DCME equipment is that it is interoperable between different manufacturers' types. The DCME utilises DSI and low rate encoding (LRE), which uses 32 kbit ADPCM involving an algorithm as defined in CCITT Rec G732 and it can service up to four destinations.

IDR specification

The full IDR specification is defined in the INTELSAT Document IESS 308 (Rev 3). In that specification the user earth station is defined both in terms of frequency modulation, phase noise, baseband characteristics, FEC and the eirp requirement when using different satellites and interconnecting between various types of earth station.

IDR eirp performance

IESS 308 contains 18 tables giving the eirp requirements for C Band, Ku Band and cross linked satellite operation at data rates from 64 kbits to 8448 kbits. Typical values of eirp are listed in Table 7.8 and Table 7.9 for C Band and Ku Band operation.

The eirp figures shown in Tables 7.8 and 7.9 are maximum eirp levels for the system in that they:

(a)   Refer to earth stations located at the beam edge;
(b)   Assume that the antennas have an elevation angle of 10°;
(c)   A receiver sensitivity variation allowance of 2 dB is included.

The eirp will decrease away from the 10° elevation angle and footprint beam edge and here certain correction factors detailed in chapter 5 must be

Table 7.8   C Band eirp requirements for range of information rates

Information rate (kbits)	Destination E/W (eirp dBW)			
64	46.8	50.1	52.2	54.4
192	51.6	55.1	57.0	59.2
384	54.6	58.1	60.0	62.2
1544	60.6	64.1	66.0	68.2
2048	61.8	65.4	67.3	69.4
6312	66.7	70.2	72.1	74.3
8448	68.0	71.5	73.4	75.6

Table 7.9   Ku Band eirp requirements for range of information rates

Information rate (kbits)	Destination E/S (eirp)		
	C	E3	E2
64	51.2	54.6	59.6
192	56.0	59.4	64.1
384	59.0	62.4	67.1
1544	65.0	68.5	73.2
2048	66.2	69.7	74.4
6312	71.1	74.6	79.3
8448	72.4	75.8	80.6

applied. In addition to the above factors the eirp figures take account of propagation changes such as rain attenuation and depolarisation effects.

The tables show recommended data rates and assume that $\frac{3}{4}$ Rate FEC is applied to the transmissions. Obviously the user may need to use other data rates and in those cases the eirp will have to be calculated from equation 7.14.

$$\text{eirp} = X + 10 \log I_r \text{ dBW} \qquad (7.7)$$

where X = value for recommended information rate dBW

I_r = required information rate

The eirp requirement can vary considerably and although IESS 308 quotes a mandatory 15 dB power adjustment range it is likely that the earth station operator will have more than this in order to cover the full range required and so retain full flexibility, whilst still maintaining an eirp level stability of 0.5 dB for Standards A, B, C and F3 and ± 1.5 dB for E3, E2 and F2.

The system recognises that adverse conditions may reduce the power arriving at the satellite and this reduction is limited to that shown in Table 7.10 for a certain percentage of time.

Table 7.10   eirp flexibility requirements for IDR satellite system

Satellite longitude	Link	M (dB)	K
Degrees East	West Up		
325.5–341.5	A	7	0.04
	B	7	0.03
	East Up		
	A	5	0.04
	B	7	0.01
307–310	East/West Up		
	A	7	0.04
	B	7	0.02
174–180	West Up		
	A	6	0.04
	B	7	0.03
	East Up		
	A	5	0.04
	B	7	0.01

where   A = 14 to 4 GHz link
B = 14 to 11/12 GHz
K = Percentage time of one year
M = Up path margin

Emission characteristics

There are various elements that are defined by the IESS document and are defined in Table 7.11.

Carrier frequency tolerance

The lifetime tolerances are listed below:

Transmit RF frequency	< ± 3.5 kHz
Receive RF frequency	< ± 3.5 kHz
Satellite translation frequency	< ± 25 kHz
	± 42 kHz

The monthly tolerance is one-tenth of the lifetime figure. The tolerances must be held to ensure the adjacent channel interference does not occur.

Phase noise

This parameter is defined in two ways:

(a) The noise is assumed to have a non-definable spectral shape and

Table 7.11   IDR emission characteristics

Type of distortion	Parameter value
Spurious noise (out of band)	< 4 dBW in any 4 kHz band
Spurious noise (in band) < 2 Mbits	> 40 dB down on carrier/4 kHz
> 2 Mbits	> 50 dB down on carrier/4 kHz
Intermodulation 6 GHz hemi/zone	$(21 - K_1)$ dBW/4 kHz band
6 GHz global	$(24 - K_2)$ dBW/4 kHz band
6 GHz global spade	$(21 - K_2)$ dBW/4 kHz band
14 GHz	$(10 - K_1)$ dBW/4 kHz band
Spectral regrowth	> − 26 dB down on main lobe
Modulation transfer	− 73 dBmOp

where $K_1 = 0.02(A_t - 10) + B_u + d\,[0.02(A_r - 10) + B_d]$
$K_2 = K_2 = 0.06(A_t - 10) + 1.0 + d\,[0.06(A_r - 10) + 2.0]$
A_t = transmit E/S elevation angle
A_r = receive E/S elevation angle
b_u = difference between satellite Rx beam edge and gain in direction of Tx E/S
B_d = difference between satellite Tx beam edge and gain in direction of worst location Rx E/S
d = fraction down link factor

Note: The full details of the adjustment factors are to be found in IESS 402.

therefore the sum of the phase noise for a single sideband should not be more than 2° or 2.8° for the double sideband case.

(b) The noise is assumed to have both a continuous and a spurious component. The continuous component must meet the envelope defined in Table 7.12 and the sum of the spurious components must be more than the 36 dB down relative to the carrier level.

Table 7.12   IDR phase noise requirements

Frequency from fc Hz	SSB noise density dBc/Hz
10	− 30
100	− 60
100 K	− 90
1000 K	− 90

Transmission parameters

The basic transmission parameters of the IDR system are shown in Table 7.13.

It should be noted that there are various data rates to be considered:

● Information rates, IR, which are the standard bit rates;

Table 7.13   IDR transmission parameters

Parameter	Value
Information rate	64–44 736 kbits
FEC < 10 Mbits	Rate $\frac{3}{4}$ convolutional coding
	Viterbi decoding
Modulation	4 phase coherent PSK
Scrambling	As CCITT Rec V35
Alarms/ESC overhead	64–384 kbits/zero
	1544–44 736 kbits/96 kbits
Min carrier BW (allocated)	0.933 (IR + OH)
	Rounded up to nearest 22.5 kHz
	increment < 10
	Rounded up to nearest 125 kHz
	increment > 10
Min noise BW (occupied)	0.8 (IR + OH)
Eb/No BER 10^{-3}	5.7 dB
$\qquad\qquad 10^{-7}$	8.7 dB
$\qquad\qquad 10^{-8}$	9.2 dB
C/T (operating point)	219.9 + 10 log (IR + OH) dBW/°K
C/N (noise BW)	9.7 dB
C/T at threshold	− 222.9 + 10 log (IR + OH) dBW/°K
C/n in BWn	6.7 dB
Threshold bit error rate	10^{-3}

- Data rates which are the information rates plus overhead;
- Transmission rates which are the bit rates after FEC has been applied.

IDR channel unit implementation

The IDR channel unit consists of a number of individual modules:

- Modulator/demodulator (modem);
- FEC encoder/decoder (codec);
- Scrambler/descrambler;
- Overhead framing unit.

Modulator/demodulator

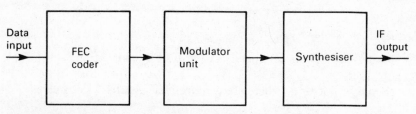

Fig. 7.4 IDR modulator.

The basic modulator is shown in Fig. 7.4. It receives the output from the FEC Encoder which is parallel data streams, designated as the P and Q channel, which have the relationship shown in Table 7.14. The modulator output must be accurate to 2°.

Table 7.14   IDR modem phase states for each transmitted bit

Transmitted bits Channel		Resultant phase
P	Q	
1	1	0°
0	1	+ 90°
0	0	+ 180°
1	0	+ 270° (− 90°)

The filter specified is a six-pole Butterworth (BTs = 1.0), with sinc^{-1} compensation as in equation 7.8.

$$\text{Sinc}^{-1} = \frac{\pi(f_c - f)}{\sin[\pi(f_c - f)]T_s} \tag{7.8}$$

and B = 3 dB bandwidth of filter

    f_c = centre frequency

    f = actual frequency

    T_s = transmissions rates in bits/sec

The above filter was chosen after a great deal of investigation by INTEL-SAT as the means of achieving the required Nyquist criteria and so avoid inter-symbol interference and provide good performance under large HPA back off conditions. It should be noticed that this area of modem performance has been relaxed in practice and can now be realised by the use of a square root 40% cosine roll off filter characteristic.

The demodulator is in a coherent QPSK form; it has to recover bit timing and the unmodulated carrier; the basic elements needed in the demodulator are shown in Fig. 7.5.

The IF filter can have a centre frequency of 70 MHz or 140 MHz, depending on the overall equipment configuration. The signal is downconverted before being applied to the four phase PSK demodulator which not only feeds out two data streams to the FEC decoder, but in addition provides an input to the carrier recovery circuitry and clock recovery system.

The circuitry also partially resolves the phase ambiguity inherent in the PSK demodulator process and it does this by feedback from the Viterbi

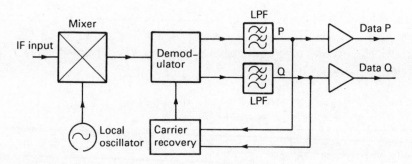

Fig. 7.5 IDR demodulator.

decoder which modifies the behaviour of the demodulator circuitry. The circuitry also performs a very important function in that it removes the frequency errors of the system by applying an automatic frequency control system that compensates for satellite, downconverter and transmitter errors, as well as Doppler effects.

Codec

The codec system is defined in considerable detail in IESS 308. The decoder receives P and Q channel signals from the demodulator which is three bit soft decision data as well as the recovered clock. The coder, which must produce a $\frac{3}{4}$ Rate convolutionally coded signal, is in the form shown in Fig. 7.6 producing a punctured form of convolutional code which is actually derived from a $\frac{1}{2}$ Rate encoder which has specific bits deleted from the $\frac{1}{2}$ Rate encoder output.

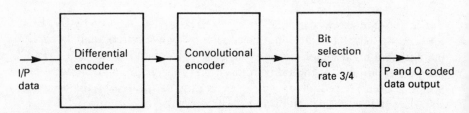

Fig. 7.6 Convolutional encoder.

The input data is first differentially encoded by a binary encoder, because the code is transparent to 180° phase ambiguity. The output from this is applied to a seven-stage register which produces the half rate encoded data; this data is then applied to a circuit which deletes two specific bits from among six bits contained in three consecutive blocks of the half rate encoded signal and those bits are encoded in the following manner to produce the required $\frac{3}{4}$ Rate encoded signal; details are shown in Table 7.15.

Table 7.15   IDR codec $\frac{3}{4}$ Rate encoding

Block	Bits transmitted/deleted
1	Both
2	Polonomial 133 Tx
	Polonomial 171 deleted
3	Polonomial 133 deleted
	Polonomial 171 Tx

The decoding is performed by a soft decision, maximum likelihood, Viterbi decoder. The INTELSAT specification sets out four mandatory requirements for the decoder:

(a)   The coding gain should match the required Eb/No;
(b)   The decoder must accept an eight-level soft decision data;
(c)   Phase ambiguity must be resolved and code synchronisation provided;
(d)   The serial output data stream must be differentially decoded.

The code used was chosen after comparing a variety of codes, either new or in use in current systems such as TDMA or IBS. If the performance is compared in terms of the number of channels that can be transmitted in a 72 MHz transponder, Table 7.16 shows the coding performance for various earth stations types.

Table 7.16   Effect of figure of merit on transmission capacity

G/T (dB/°K)	Number of 64 kbit channels				
	none	7/8 HD	7/8 SD	3/4 SD	1/2 SD
29	200	460	600	720	800
32	230	580	800	1000	800
35	350	810	1050	1200	800
38	500	1100	1360	1200	800
41	610	1400	1400	1200	800

The final choice was made on the basis of the Standard A and Standard C earth stations.

Signal scrambling
This is performed for the same reason that energy dispersal is applied to analogue TV systems and that is to disperse the signal over the full bandwidth and thus avoid peaks of transmitted power that will take the system outside the specification laid down in CCIR Rec 358–3 as well as meeting the off axis performance defined by CCIR Rec 524–2. The scrambling method and performance are defined in CCITT Rec V35, which must be placed

before the encoder and after the decoder i.e. not inside the coding/decoding process as the scrambler is an error multiplier and can produce three errors for every single real error over a bit length of 20 bits. The scrambler however has the advantage not only of simplicity in its realisation but also it is in use in existing modems.

Overhead framing
The information rate data streams are required to carry further information that does not add to the quality of the demodulator output but is required for two purposes:

- To provide an engineering services circuit (ESC);
- To carry maintenance alarms.

Table 7.17 shows the additional bit rates that are added to the information rate. The way in which these are added is defined for CCITT standard data rates. The method of adding the overhead is simple, i.e. 12 bits in each 125 microseconds-frame-period up to 6312 kbits and 16 bits for 8448 kbits; the bits are used as shown.

Table 7.17   Overhead data required for ESC and maintenance functions

Function	Data rate used (kbits)
Frame alignment	20
Backward alarms	4
ESC data	8
ESC voice	32
ESC voice	32

The faults that are covered by the system are:

- Failure of equipment in earth station resulting in prompt alarm;
- Loss of signal from terrestrial circuit resulting in AIS to terrestrial systems;
- Loss of incoming signal from the satellite resulting in AIS to satellite and terrestrial prompt alarm in station;
- Loss of overhead frame alignment resulting in AIS to satellite and terrestrial prompt alarm in station;
- Loss of overhead multiframe alignment resulting in AIS to satellite and terrestrial prompt alarm in station;
- BER exceeded over one-minute period;
- Distant earth station alarm.

IDR channel unit performance

Bit error rate performance
The channel units must perform in a degraded environment in which there are:

- Adjacent carriers at a level of $+7$ dB up on the carrier;
- Frequency errors on the carrier frequency;
- Adjacent carrier frequency errors.

The system may also produce errors, both in the scrambler and FEC system and therefore on a back to back measurement at IF these must be assumed to be in the circuit. Table 7.18 gives the performance required.

Table 7.18   IDR BER performance

BER	Eb/No (dB)
10^{-3}	5.3
10^{-7}	8.3
10^{-8}	8.8

(Eb/No refers to the modulated carrier and the figures assume that the data entering the FEC coder is at the data rate as defined previously.)

Transmission delay/timing
The existence of signal delay in satellite communications and the variations of that due to satellite movement have been dealt with earlier; this variation has the effect of producing errors in the link performance due to the Doppler shift effect.

The other element that must be considered is the timing of the various signals, timing which can be derived from:

- An atomic reference standard;
- A reference timing such as Loran C;
- Terrestrial PCM clocks when analogue voice channels are involved;
- A clock signal transmitted from a remote earth station, which itself derives its reference from a local reference standard.

The system utilises data buffers to avoid delay and timing problems, with the size of the buffer being defined by INTELSAT for three types of network connection with various satellite inclinations. The figures shown in

Table 7.19 Channel unit buffer size requirements

Satellite	Transmission delay (max)	Rate of variation of transmission delay
5, 5A, 5A (IBS)	0.6 ms	18 ns/s
6	0.34 ms	9.7 ns/s

Table 7.19 relate satellite inclination to buffer capacity for the three connections which are:

(a) Transmission from digital to digital networks using atomic standards;
(b) Timing derived from receiving earth station demodulators and retransmitting to the originating earth stations;
(c) Timing derived from receiving earth station demodulators and transmitted to another earth station.

The atomic standards needed to give an accuracy of 1 part in 10^{11} are most used in national digital networks and then at the most used hierarchy i.e. the primary level of 2048 kbits or 1544 kbits. It is for this reason that buffering takes place at this level even for higher order transmissions up to 8448 kbits, which must perform its buffering when the signal has been demultiplexed down to the primary level.

Baseband characteristics
The primary reason for the IDR service is to allow satellite earth station operators to insert themselves into the terrestrial digital networks and so appear to be transparent to these systems. This requirement means that it is expected that the traffic should be carried on CCITT Rec G802 hierarchies and therefore the use of any other carrier sizes must be individually agreed with INTELSAT.

The need to interface with terrestrial digital networks not only requires conformance with G802 but will also need to define the hierarchy being used. Table 7.20 shows the IDR multiplex characteristics for multi-destination operation; this defining process will ensure that the treatment of alarms is the same for terrestrial and satellite networks, though for multi-destination operation there are still ambiguities when a fault condition is sent to all destinations.

One possible method of overcoming the phase ambiguity is to limit the number of destinations as is done in IDR alarm philosophy; another method is to use DCME with the supervisory information being carried in the time slot 16 of the 64 kbit channel. The IDR network is still in the process of development and no doubt will change its parameters over a period of time to take account of the changing customer requirements.

Table 7.20 CCITT recommendations for IDR transmission rates

Transmitted Bit Rate (Mbits)	CCITT mux recommendation
1544	G733
2048	G732
6312	G747
8448	G742
32064	G752
34638	G751
44736	G752

7.4.4 Earth station design

The design of the earth station for the IBS service can be computerised in order not only to achieve the optimum performance at the least cost, but also to ensure that the repetitive activity of system design does not add to the cost of the earth station. The computer takes account of the electrical parameters and also the actual position of the two earth stations that are communicating with one another. The necessary eirp is listed in the INTELSAT document IESS 309 giving the parameters for two grades of service:

Basic IBS

This is designed to maximise the channel capacity in both C and Ku Band operation, and essentially gives a better availability to the C Band system than the Ku Band.

Super IBS

By increasing the eirp the availability of the Ku Band service can be increased to match that at C Band.

The figure of merit for the various earth stations has already been defined and this G/T can be realised in many ways and a performance/cost trade-off must be made.

The first consideration to be made is in the antenna used. The system gain is a direct function of the antenna diameter and therefore the G/T is also directly proportional to it. The minimum cost antenna will be one that does not have a tracking system, for this means that the antenna can be simpler as it will not need a tracking receiver, beacon downconverter, antenna control system or any form of actuator or motors to drive the antenna to a new position. For the INTELSAT business systems the use of tracking is not necessary for antenna diameters up to 3.7 m, while for diameters up to 5.6 m tracking is not necessary provided that manual adjustments are made on a daily basis. The reduction in the antenna diameter results in an adjustment in two other parameters:

(a) The LNA must have a reduced noise temperature. This can be an advantage if a GaAs FET LNA is used but is not cost effective if the system reliability requires that a redundant low noise front end must be used.

(b) The traffic capacity must be reduced or alternatively the HPA output power must be increased consistent with the off-axis power emission limitations. The eirp increase, to maintain a particular traffic capacity, involves an increase in one of the most expensive elements in the earth station. Even if this can be done in a cost-effective manner for small antennas the off-axis eirp will exceed the specified limit. For instance if communication is between two 3.7 m antennas then the off-axis limits will be exceeded.

One method of improving the off-axis performance is to design an antenna that has a higher gain for a given diameter with an improved sidelobe performance, as obtained by use of offset antenna geometry. This area of design has shown considerable advances in recent years and we now see antennas with a gain of 46 dB, equivalent to a circular parabolic antenna of 2.2 m. Design has led to an effective sidelobe response improved by up to 10 dB, thus making it possible to transmit higher eirp without exceeding the off-axis emission characteristics. The relationship between the antenna size, LNA and figure of merit is shown in Tables 7.21 and 7.22, which cater for both C Band and Ku Band as well as each type of terminal.

Table 7.21   IBS C Band E/S figure of merit for range of antenna sizes

Antenna diameter	3.5	6.1	9.0	13	18
Gr	42	49.5	53.1	56.6	59.8
G/T (70°)	20.8	25.5	29	32.5	35.3
G/T (45°)	21.9	26.6	30.1	33.6	36.4

Table 7.22   IBS Ku Band E/S figure of merit for range of antenna sizes

Antenna diameter	3.5	6.1	9.0	13
Gr	50.2	55	58.5	63.7
G/T (225°)	25.0	29.8	33.7	37.1
G/T (150°)	26.3	31.1	34.6	38.0

The tables reflect the performance with the antenna at an elevation angle of 10°. For other elevation angles the figures must have a correction factor applied: in the C Band case the G/T is improved by up to 0.5 dB while the Ku Band case is only improved by up to 0.2 dB.

Earth station transmit performance

The HPA required is dependent upon the traffic capacity of the system. The eirp available for various antennas and HPA powers is shown in Table 7.23.

Table 7.23 IBS eirp performance

C Band				
Antenna diameter	3.5	6.1	9	
eirp	63.3	67.9	71.8	HPA 75 W
eirp	70.5	75.1	79.0	HPA 400 W
Ku Band				
Antenna diameter	3.5	6.1	9	
eirp	70.3	74.9	78.4	
eirp	76.1	80.7	84.2	

The eirp is assumed to be operating with a saturated HPA and it has to be recognised that it is necessary to operate with the HPA output backed off from saturation, normally by 3 dB if only a single carrier is being transmitted, but in the case of multi carrier operation this back-off must be increased to 7 dB for C Band and 89.5 db for Ku Band operation. The design is further complicated by the fact that the back-off is set for carriers of equal size, whereas a practical system will not have equal carriers and therefore the back-off can be reduced, the amount depending upon the mix of carriers.

The design parameters are to be found in IESS 309 but practical examples are shown in Tables 7.24 and 7.25.

Table 7.25 applies to a Ku Band system that has an LNA noise temperature of 225° and an antenna size of 3.5 m (6.1 m). The assumptions made are that the earth station operates in a multi-carrier mode, with a back-off of 3 dB for a single carrier and 7 dB for multiple carriers.

One of the most impressive realisations of the INTELSAT business terminals has been accomplished by Satellite Transmission Systems (STS).

Table 7.24 64 kbit capacity for Standard E/F C Band earth stations

HPA		10	75	400	3000
Capacity	F1	0 (0)	1 (1)	2 (6)	13 (33)
	F2	0 (1)	1 (2)	2 (15)	34 (84)
	F3	0 (1)	1 (3)	8 (18)	45 (97)
Capacity	E1	0 (1)	1 (3)	6 (16)	36 (88)
	E2	0 (1)	2 (5)	12 (29)	64 (157)
	E3	1 (3)	9 (22)	49 (120)	281 (641)

Note: The figures show the number of 64 kbit carriers that can be supported, using a 6.1-m antenna and a 9-m antenna (shown in brackets).

Table 7.25   64 kbit capacity for Standard E/F Ku Band earth stations

HPA output power	2	5	80	300
Capacity F1	0 (1)	1 (1)	5 (17)	22 (64)
F2	1 (1)	3 (2)	14 (41)	53 (153)
F3	1 (1)	1 (1)	22 (63)	82 (238)
Capacity E1	1 (1)	2 (2)	16 (47)	61 (176)
E2	2 (2)	1 (5)	31 (89)	116 (336)
E3	3 (9)	8 (23)	129 (374)	486 (700)

Note: The figures show the number of 64 kbit carriers that can be supported, using a 6.1-m antenna and a 9-m antenna (shown in brackets).

They have produced a very flexible equipment system that is called a Self Contained Antenna Mounted Package (SCAMP). In practice the earth station format is the same as any other supplier's but it is designed as an integrated system. The design follows a logical sequence in that it is modular, with the whole of the HPA and GCE mounted on the antenna hub. These are also modular and can be added to as required, to increase the earth station capacity. The actual performance of the terminal allows operation on a worldwide basis, being able to vary the Ku Band front-end to give the correct frequency band. These bands are given in Table 7.26.

Table 7.26   Worldwide Ku Band frequency bands

Region	11 GHz	12 GHz
N/S America	10.95–11.2/11.45–11.7	11.7–11.95
Europe/Africa	10.95–11.2/11.45–11.7	12.5–12.75
Asia/Australia	10.95–11.2/11.45–11.7	12.5–12.75

The earth station has also been fitted with a monitoring and control system which is microprocessor-based and allows the operator to control the system either remotely or locally.

7.5   EUTELSAT BUSINESS SERVICES

7.5.1   Introduction

The EUTELSAT organisation established a satellite business service known as Satellite Multiservice System (SMS). This service is made up of two forms of network, the difference between the two is the method of transmission, with one using a TDMA form of access that is a fully demand assigned system, while the other uses an SCPC FDMA access method. The purpose of the networks is very similar to the INTELSAT system in that it provides a small antenna service to the business user though obviously on a restricted basis to European countries only, with the same type of services on offer such as voice, data, text and compressed video. It also follows the INTELSAT pattern in that it offers both open and closed network operation – requiring different standards to be met by the user.

The first phase of the EUTELSAT system not only used its own satellites but took capacity from the French TELECOM 1 system. The EUTELSAT 2 series of satellites however, will separate the two networks.

7.5.2 TDMA network

The original TDMA system used Ku Band for both transmission and reception, operating at 24.576 Mbits, using a 2–4 phase shift keying modulation which is in reality bi-phase shift keying which utilises less spectrum.

As already stated the EUTELSAT 1 TDMA system was carried by the French TELECOM 1 satellite but for the EUTELSAT 2 system the service will be carried on its own satellite, at the same time it will also carry the next generation of TELECOM TDMA service. The new satellite will operate with the existing TDMA earth stations because the technical parameters remain the same between systems. The basic parameters of the TDMA system are:

Modulation	2–4 phase PSK
Demodulation	Differential
Encoding	Differential
Bit rate	24.576 Mbits
Multiframe	1.28 ms
Frame length	20 ms

Once again the transponder hopping is accomplished at the 140 MHz point. The opportunity was taken in designing the system performance for EUTELSAT 2 to improve the antenna performance to ensure that dual polarisation can be used. To do this the cross polar isolation has been increased to 35 dB. It is intended that dual polarisation systems will be employed even though, for low traffic capacity, it is perfectly efficient to operate on separate polarisations for transmit and receive. The antenna performance is detailed in Table 7.27.

The quality of the service provided by the TDMA system must be sufficient to meet the CCIT Recommendation 614 shown below:

Table 7.27 SMS TDMA station antenna parameters

Parameter	Value	
Rx Frequency range	12.5–12.75 GHz	
Tx Frequency range	14.0–14.25 GHz	
Sidelobe/mandatory	$32-25 \log \theta$	$1° < \theta < 48°$
	-10	$\theta > 48°$
Sidelobe/target	$29-25 \log \theta$	$1° < \theta < 20°$

BER	Percentage of time
10^{-7}	10
10^{-6}	2
10^{-3}	0.2

This performance is worse than the current EUTELSAT 1 satellites but the new performance can be met over the whole of the footprint under all forms of environmental conditions. The sidelobe performance required follows the CCIR Recommendation 580, which stipulates that the target shown above should apply to all antennas built after 1987. This recommendation did not apply to antennas that had a d/l less than 150, i.e. below 3.2 m at 14 GHz. Now this means that for the EUTELSAT 1 earth stations the antennas may not meet the target specification. However this can be alleviated by using later designs for the EUTELSAT 2 earth stations when the smallest size is applicable and the earlier antennas for other applications.

7.5.3 TDMA earth station performance

The earth station is specified in terms of figure of merit, eirp and bit error rate performance. In the EUTELSAT 2 design the designer must take account of the general need to ensure that the system is as simple as possible and can be implemented in a very cost-effective manner and thus increase the service take-up. The overall satellite system design carried out by EUTELSAT is deliberately tailored to allow the earth station designer to use antennas that do not require tracking systems and are therefore simpler. The figure of merit is given in equation 7.9.

$$\frac{G}{T} = 27 + 20 \log \left(\frac{f}{12.5} \right) \text{ dB/°K} \tag{7.9}$$

where f = frequency in GHz

The eirp is specified as:

$$E = 73 + 20 \log \left(\frac{f}{14.0}\right) \text{dBW} \qquad (7.10)$$

The parameters of a typical earth station are given in Table 7.28.

Table 7.28   TDMA earth station element parameters

Element	Parameter value
Antenna diameter	3.5 m
LNA	120 °K
HPA	200 W/400 W
Pointing loss	0.7 dB
Loss HPA to feed	1.5 dB
Satellite G/T	2.0 dB/°K
Satellite eirp	44.0 dBW

The antenna will be non-tracking, dual polarised with a typical receive gain of 45 dB and a transmit gain of 52.5 dB. The LNA design assumed that there would be an equipment capable of providing a noise temperature of 120°, an improvement of 180° compared with the EUTELSAT 1 system. The HPA figure recognises that there are different coverage areas to cater for; the 200 W will provide a service for most of Western Europe; the 400 W amplifier will increase that coverage to Portugal and parts of Turkey.

7.5.4   Satellite multiservices system (SMS)

In the EUTELSAT system the business service specifies open network parameters that the user has to meet in order to access the system. These parameters are the same between EUTELSAT 1 and 2 satellites, though the earth station realisation is different due to the different satellite character- istics. The EUTELSAT system follows that of INTELSAT in providing both an open and closed network for the user to operate in. The closed network is extremely flexible in that although it is still based on the use of the same satellite it can provide a communication service which can utilise a variety of modulation and accessing methods. The second method of transmission for the open network is to use FDMA/SCPC which is directed to the small capacity user. The ECS 2 satellite will use two types of transponder provid- ing either 36 MHz or 72 MHz bandwidth, the latter being used when the earth station operates with a 140 MHz IF frequency. The general parameters of the system are given below.

Figure of merit G/T

The second generation of satellites specifies three forms of earth station as listed in Table 7.29.

Table 7.29   SMS earth stations figure of merit

Earth Station Standard	G/T db/°K
1	$29.9 + 20 \log (f/12.5)$
2	$26.9 + 20 \log (f/12.5)$
3	$23.3 + 20 \log (f/12.5)$

The earth station can be realised in a variety of ways depending upon the trade off between antenna size, LNA noise temperature and HPA size. Typical earth station parameters are shown in Table 7.30.

Table 7.30   SMS earth station parameters

Earth station equipment	Standard		
	1	2	3
Antenna diameter (m)	5.5	3.7	2.4
LNA °K	180	180	180
HPA Power/W 64 kbits	3.5	5.3	12
2.64 kbits	10	15	34
2048 kbits	52	78	178

Effective isotropic radiated power

The eirp differs between the Series 1 and Series 2 satellites, EUTELSAT 1 requiring 55.3 dBW for a 64 kbits carrier and EUTELSAT 2 requiring only 52 dBW. The HPA output back off is determined by the sidelobe level compared to the nominal carrier level; for EUTELSAT 2 this has to be 4 dB.

Antenna characteristics

The antenna design must meet the same parameters as those described in the TDMA section. The off-axis eirp requirement is set by defining the amount of power in a 40 kHz band at particular off axis angles.

The design parameters are shown in Table 7.31.

Table 7.31   SMS antenna characteristics

Pointing angle	eirp
$2.5° < \theta < 20°$	$36 – 25 \log \theta$ dBW
$20° < \theta < 26.3°$	3.5 dBW
$26.3° < \theta < 48°$	$39 – 25 \log \theta$ dBW
$48° < \theta$	-3 dBW

These parameters allow the operator to use a 1.8 m antenna to transmit a 64 kbits carrier.

7.5.5 EUTELSAT closed networks

The closed network can operate in TDMA, FDMA or spread spectrum modes and use a variety of modulations such as FM or some form of phase shift keying. The system parameters are detailed in Table 7.32.

Table 7.32   EUTELSAT closed network system parameters

Parameter	Value
Antenna sidelobe	As the open network
Polaristion discrimination	35 dB
Pointing stability	0.3 dB
Frequency band	As the open network
eirp stability	0.5 dB
Out of band emissions	-55 dB below carrier level
Scrambling	To CCITT Rec V35

The trend to smaller earth stations means that the communication networks are becoming more personal and the EUTELSAT system attempts to address this VSAT market; this will be dealt with in detail in chapter 8.

7.6   NORTH AMERICAN BUSINESS SYSTEMS

The USA market is immensely diverse and produces most of the innovations that have arisen in satellite communications. The earth station characteristics differ only due to the fact that the North American frequency allocations differ and the data rates for any transmission use a different format. This difference produces considerable difficulty in the interconnection between the USA and Europe.

One major area of difference in earth station design is in the antenna definition, where the parameters are set by the FCC. For the USA market the antenna must meet the specification shown in Table 7.33.

Table 7.33   FCC antenna specification

Pointing angle	Antenna gain pattern
$1° < \theta < 7°$	$29-25 \log \theta$ dB
$7° < \theta < 9.2°$	8 dBi
$9.2° < \theta < 48°$	$32-25 \log \theta$ dBW

This specification also defines the amount by which a single sidelobe can exceed the performance necessary; for the first region of pointing angle the single lobe cannot be more than 3 dB above the specification; for the other areas of antenna pointing the limit is set at 10%. This specification is tighter than in the first region but easier to meet at other pointing angles. It should be noted that the tighter specifications are difficult to meet and this has given a strong impetus to off-set Gregorian antenna design which has been shown to have a much better sidelobe performance potential than the standard symmetrical antenna.

Chapter 8

Very Small Aperture Terminals (VSAT)

8.1 INTRODUCTION

The historical trend towards smaller earth stations, due to higher power satellites, better earth station design and a deregulated environment, has resulted in the development of very small terminals that can be situated directly on a user's premises to provide data and video links. The subject warrants a chapter in its own right because of its increasing importance in the field of satellite communications.

The introduction of very small aperture terminals or VSAT systems has been both technology and commercially-led. The term VSAT refers to earth stations that use antennas that have a diameter of less than 1.8 m. The accelerating trend towards digital communications and the growing importance of the financial and business sector to the communications market has produced solutions that could only have been achieved by improvements in satellites and satellite launchers, more cost-effective earth stations, subsystems and the development of the use of the higher frequency bands at 11–14 GHz (Ku Band).

The first major application of such terminals was very much the result of one company's efforts, i.e. Equatorial founded in the USA in 1979 with the intention of developing a satellite system that would convey data between users' premises. The service was seen as essentially low speed, compared to other satellite systems or optical fibre but this was seen as a major area in which the VSAT could operate.

The Equatorial VSATs operated at C Band, which they considered to be more cost-effective than Ku Band and less susceptible to such problems as rain attenuation and fading. The initial earth stations were receive only and used spread spectrum techniques as a transmission method. These earth stations were:

- Very low in cost;
- Simple to operate;
- Physically very convenient;
- Highly reliable, giving high availability;
- Essentially interference-free and therefore able to operate in a crowded electronic environment.

The huge success of Equatorial, which has an installed base of thousands of terminals and now has its own satellite capacity, encouraged a large number of suppliers into the market, mainly operating at Ku Band, with a proliferation of transmission methods and protocols and a wider range of facilities for both one and two-way communications.

A typical VSAT system is shown in Fig. 8.1, which is using the central hub concept; in this the VSATs are used as the remote terminals and are connected together via the hub station. The use of a hub station is crucial in a VSAT system in that it removes the need to design stations with the ability to communicate with any other VSAT.

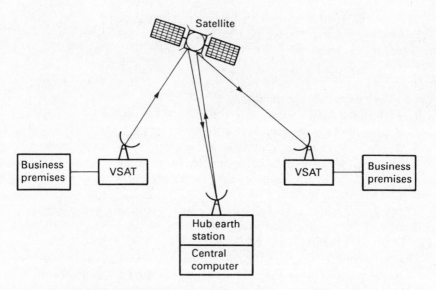

Fig. 8.1 Typical VSAT star network.

The two-way VSAT system must have a high-level signal from the satellite, to allow the use of the small antenna, implying that the eirp transmitted up to the satellite is high. This obviously could not be from another VSAT as it would have low antenna gain and small HPAs. It is therefore necessary to use another, larger earth station as its power translator. In the uplink mode the hub station needs less power from the VSAT due to its own large antenna; in the downlink mode the hub station can send a high eirp to the satellite and thus reduce the VSAT size required for good reception. The star system of operation is the major one used for VSAT networks, with the advantages already stated. The drawback to the system is the fact that to connect between other VSATs requires a double hop with all the problems associated with delay due to transmission distance.

8.2 VSAT APPLICATIONS

The range of uses available to the VSAT user depends upon the size and complexity of the terminal involved. Obviously if the number and range of services are increased to include high-speed data, two-way video etc., the terminals begin to increase in size and complexity until they resemble the range of INTELSAT Standard E and Standard F terminals, as described earlier in chapter 7. The application is for data transmission and whatever the original information, be it voice, video, fax etc. then that information has to be digitalised at the appropriate bit rate before transmission.

The first division of VSAT usage is into one or two-way services, implying the addition of HPAs to the receive terminal. Typical non-interactive services are:

- Video links to stores, garages, etc. to display advertising information;
- Education programmes to remote sites, as in India and Australia;
- Ticker tape type information that updates financial offices;
- Inter-office information on organisational changes etc.

If the VSAT is upgraded to work in a duplex mode then the usage expands considerably:

- Computer to computer data interchange and database interrogation;
- Video-conferencing, either within companies or between businesses;
- Point of sales bank account checking etc.;
- Airline/car reservations systems;
- Remote process control;
- Financial transactions;
- Two-way educational, i.e. interactive teaching.

The area of use is very wide indeed as VSAT networks can provide the user with a complete system which can be used to provide the facilities that a local carrier or PTT cannot. The user thus obtains a total independence from the indigenous carrier and so develops the system as necessary within their own control. This freedom of action is a very attractive feature to the corporate user, bringing as it does flexible communications, matching individual requirements and which can be installed in a very short space of time.

Initially the most widely used function of a VSAT network was in one-way point to multipoint communications but this restricted use is more and more giving way to a duplex service which can distribute data on an interactive basis. It allows, for instance, the insurance industry to maintain an immediate awareness of the activity of its field agents, without the necessity for travel to a central or regional office, or for that matter the need to communicate via the public telephone network. In this particular area of application it can be seen that the VSAT can provide a method by which more people can work from a home environment by simply installing a terminal on the worker's premises. This not only reduces the length of the working day, but reduces the road traffic and other loads on the community as well as a company's travelling expense budget, while allowing the worker to operate in an environment that gives them full access to company data etc. This facility can be improved even further by the addition to the VSAT of two-way video thus allowing meetings to be held between remote locations.

In essence the VSAT is very suitable for all areas of activity where information can be transmitted, without error, between remote locations. The VSAT can thus revolutionise retailing, both at the customer level and stock control. It can provide the facility for training and education from a centralised source, it can be used to organise the despatching of goods and the co-ordination of shipments. It can, by reason of the improved communications, speed up the decision making process and disseminate those decisions to the workforce much more quickly and reliably.

8.3   VSAT NETWORK CONFIGURATIONS

The network does not have to be connected in a star arrangement and will depend upon the users' requirements with consequent implications for system and equipment design. There are basically three forms of network configuration: star networks, broadcast networks and total node connection networks.

8.3.1   Star network

This has already been mentioned and utilises a central hub station which provides the interconnection between all the remote terminals, which cannot communicate directly with one another. A typical form is shown in Fig. 8.1. If there are n nodes in the network then in this case there are $n - 1$ trunk groups to be catered for.

8.3.2   Broadcast networks

This is, in one sense, a special case of the star system because the information is always routed between the central hub and the remote station, but is non-directional between the remote terminal and the central hub earth station. This is particularly applicable where the network is used to relay information to remote sites that do not need to be interactive.

8.3.3 Total node connection network

In many networks the interconnections pattern is as shown in Fig. 8.2, where

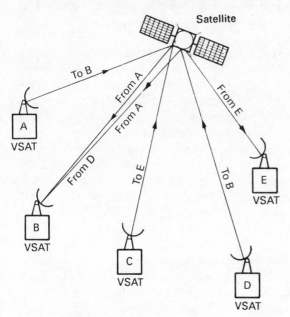

Fig. 8.2 VSAT system with total node connection.

it is more complex and each remote terminal would require to be flexible, with the possibility of communicating with any other terminal. It is particularly applicable to voice networks, though with the integration of voice and data transmissions the network designer will be increasingly unable to differentiate between the two requirements and separate voice from data.

8.4 VSAT SYSTEM DESIGN

8.4.1 Introduction

The design of a VSAT network is dependent on a number of factors:

- The transmission bit rate required;
- The number of remote terminals;
- The throughput of the system;
- The facilities required on the service;
- The frequencies and satellites available;
- The central control requirements.

8.4.2 Satellite availability

The system designer will have the choice of two major satellite bands in some areas of the world, such as Europe or North America but in many areas, vhere the satellite is part of the country's domestic system, such as India, here will only be a restricted choice.

In general the designer will be able to use C Band in the 4–6 GHz range or Ku Band in the 11–14 GHz range. The increasing use of Ku Band came about as a result of overcrowding in C Band and the reduction of satellite spacing to 2° of the orbital arc. This implies that at C Band the system will be operating in an interfering environment. This interference has meant that in many VSAT manufacturers' systems the technique of spread spectrum is used, a technique that overcomes the interference problem to a large degree, though at the cost of requiring a much larger bandwidth. It is in fact the most inefficient form of access method in terms of number of channels accommodated in a defined bandwidth. The Equatorial system was the first to use spread spectrum techniques and due to its patent protection ensured that its window of opportunity in the market place was extended considerably.

The access method used by Equatorial was known as Code Division Multiple Access (CDMA), in which the data is superimposed onto a specially coded waveform. This composite signal is used to modulate an RF carrier before transmission. The spectrum of the transmitted signal is therefore very broad, occupying most of the transponder bandwidth. The RF carrier frequency can thus be the same for each VSAT terminal, with each terminal having its own unique modulation identity which the VSAT receiver can recognise. The receiver will essentially be a cross-correlation system that will only respond to its programmed identity, allowing the transmit spectrums to overlap without interfering with one another.

Another form of modulation/access is one commonly used in larger earth stations, i.e. FDMA/SCPC, which is perhaps the simplest to implement. The use of SCPC ensures that a number of carriers can be transmitted without producing intermodulation distortion or interference with one another. However this method is not very cost-effective for low data rates, where a large number of VSATs, say 100, are involved as each VSAT would require 100 modulators and demodulators. One method by which the SCPC system can be made more efficient is to provide a computer control system at the hub station that can make the network operate as a demand assignment multiple access (DAMA) system, thus ensuring that the SCPC channel is only allocated as required, not as a fixed pre-assigned channel.

Digital communications has led to another form of access system, i.e. time division multiple access (TDMA), described in detail in chapter 3. This is the method that allows a number of earth stations to share a common satellite transponder. While the method involves a relatively complex receiver, with a consequent increase in cost, the overall system is easy to re-configure and increase in size as necessary.

The general operation of VSATs is of an intermittent nature, at relatively low bit rate, even for video conferencing. This characteristic means that the choice of TDMA is best realised in a random access form rather than as a fixed time slot allocation. The control of the overall system, of time slot allocation and control is performed by a central computer situated at the hub station. The remote terminals are allocated a specific return channel, with a number of terminals having the same return channel but each channel has a number of packet frames, with any of the terminals able to access any packet. This access method means that any packet may collide and if this happens then the VSAT terminal will sense that it has happened and after a random period of time, within half a frame interval, the packet is re-transmitted, with little chance of further collision.

8.4.3 Satellite frequency

The two major frequency bands, C and Ku Band, already mentioned, are used for VSAT operation but the major systems use Ku Band for a variety of reasons:

(a) The antenna characteristics are such that the antenna beam is much narrower due to the frequency being almost three times that at C Band. This means that the interference problem, vis-à-vis satellites at 2 degrees spacing on the orbital arc, is very much reduced.

(b) The Ku Band operating frequency band means that the possibility of terrestrial interference is less, as terrestrial communications operate mainly at other frequency bands, whereas the use of C Band brings the operator directly into one of the most crowded frequency bands, i.e. 6 GHz. This characteristic is very significant as it means that VSATs can be used without involving the controlling authority in the complexity of co-ordination procedures. This is especially important in the USA.

(c) The increase in frequency does not mean that there is an advantage in antenna size reduction; this is offset by the increase in the up and down path attenuation.

(d) The disadvantage of moving to a higher frequency, i.e. increased atmospheric attenuation and depolarisation effects, can be offset by allowance being made in the system planning and also antenna design that can counteract the depolarisation effects, at least to some extent.

8.4.4   Transmission bit rate

The bit rate will depend on the services required and within that requirement whether, for instance, the video transmision has to be of broadcast quality or not. The effect of the higher bit rate is to increase the complexity of the receive and transmit terminals and many VSAT providers have different terminals for different applications, with a larger antenna, typically increased from 1.2 m to 1.8 m for video applications and improved LNC systems of higher frequency and phase stability.

The satellite transponder capacity requirements will also increase with the broadcast quality video, requiring almost 36 MHz transponder bandwidth unless some form of compression system is used. This has a consequent increase in capital expenditure but reduction in long-term operating costs.

A typical terminal will provide transmit and receive capability for data rates between 1220 bits/sec to 56 kbits, while for television a data rate of 384 kbits for compressed systems and 2048 kbits for broadcast quality would be required.

A typical system will be able to support both synchronous and asynchronous data; for synchronous from 2.4 to 19.2 kbits, 56 to 896 kbits and 1.536 (2048) kbits.

For services such as electronic mail which is not a real-time service, the transmission system parameters follow the normal terrestrial voice and data services and will require the following:

- 64 kbits PCM;
- 32 kbits CVSD delta modulation.

A rapidly increasing form of communications is the facsimile service and for this a variety of data rates are required depending upon the type of system used, which is usually a function of quality of the document that has to be handled. For general use a 2.4 kbits data rate may be sufficient, but for high-speed copiers anything up to 56 kbits may be needed; once graphics are involved then the data rate could increase to 448 kbits.

The video services again require different rates depending on the quality required. Slow scan techniques are used where there is no need to show rapid motion and these can operate at 9.6 kbits; for full motion video 2048 kbits is required.

8.4.5   Central control system

The common star configuration utilises a larger earth station as a hub through which the VSATs communicate. The hub station can be specially designed for the individual network or utilise one of the INTELSAT standard gateway earth stations such as a Standard C or Standard E. The hub function will depend on the method of access used but will require equipment for the transmission and reception of data signals at a variety of bit rates.

The hub consists of:

(a) An antenna whose size depends on the size of the network, though this does not allow for system expansion and often the initial installation is made larger than necessary in order to allow for future expansion.
(b) The low noise front end of the receive system made up of a low noise converter using GaAs FET technology and with a noise temperature in the region of 180°. The LNC is followed by a downconverter that will, depending upon the system design, translate the signal down to a 70 MHz or 140 MHz intermediate frequency.
(c) In the transmit sub-system the modulated signal is fed to an upconverter which will convert the IF signal up to 6 or 14 GHz, depending upon the satellite frequency. It is then applied to an HPA that is coupled to the antenna feed.
(d) Communication sub-systems are a part of the non-attended earth station remote from the access system being used to control the overall VSAT system. For this reason different earth station equipment will provide status information and alarm conditions to a monitoring and control system which not only checks the system status but allows redundant equipment to be brought into action should circuits fail.

8.5 VSAT PROTOCOLS

8.5.1 Introduction

The essence of VSATs is flexibility, both in the ability to install terminals easily and to adapt the network to an increasing and variable traffic. The VSAT is not a high-volume constant level of data but has traffic over relatively short periods with intermittent transmissions. In general the limit to the amount of data which can be transmitted is not power-limited as the satellites use relatively high power transponders, but by system bandwidth and therefore the utilistion of bandwidth is a premium consideration. The network operator will wish to use as little bandwidth as possible in order to minimise costs. The usable bandwidth will depend upon the efficiency of the control system in allowing each remote terminal to access the system and therefore a multiple access system must be able to minimise the number of information collisions and also reduce the delay experienced by the message.

The term protocol, which was dealt with in chapter 3, is essentially a set of rules by which a number of users share the same channel of communication.

The rules determine the ease with which the system users co-exist and try to take account of the various parameters of the system, such as, activity rate, bandwidth available, allowable message delay etc.

The multiplicity of access systems has led to a very complex set of variables for each system and the designer has to consider:

- The allowable message failure rate;
- The average delay in transmitting a message;
- The maximum and minimum data rates involved in the system;
- The type of message, whether it is of fixed length, linearly variable length or exponentially distributed length.

8.5.2 Specific protocols

ALOHA protocol

The ALOHA protocol was first proposed by N. Abrahamson in 1970 as a protocol for inter-computer communications. The protocol was designed to provide a simple, low-delay method for asynchronous operation, though it had the disadvantage that the system could only deal with low capacity networks. For fixed length messages the capacity is 18% which reduces, if variable length messages are used, to 13%.

Slotted ALOHA protocol

In a slotted ALOHA system the basic ALOHA system is retained but it restricts the transmissions from the remote terminals to the start and finish of the allocated time slots. This restriction means that the chance of collision is reduced and therefore the usable system capacity can increase to 37% of the total system capacity for fixed messages, though the system complexity increases.

Selective spectrum ALOHA

The use of spread spectrum has already been discussed with respect to its ability to allow operation in an interfering environment. This property can be used to give improved performance when incorporated into a system which uses the ALOHA protocol, as packets will not be corrupted by collision due to the effect of SSMA. It does however require a forward error correction system, with a consequent increase in circuit complexity.

Time of arrival ALOHA

If messages are made a fixed length and the circuitry is arranged to analyse the collision of packages, then with unslotted operation, those collision analyses can pinpoint the first and last packets that were in a collision and thus determine the times when packets can be retransmitted. This selection gives a usable system capacity of 40 to 50%, depending upon the use of collision detection, as well as signal detection. This system gives low delay, better capacity and less circuit complexity than selective spectrum ALOHA. One variation on it is known as tree CRA which is a slotted protocol that does not have a random access method, but sets rules concerning retransmission of packets and how access is allowed. The basic method is that any terminal, whose last message was involved in a collision, partitions that message until all collisions have been resolved. The method can give a usable system capacity of 45%.

Announced retransmission random access (ARRA)

A high capacity system can be achieved by the use of an information tag on the transmitted message. This tells the system which slot will be used for retransmission of that particular message, thus reducing the possibility of collision. The advantage of this method, which achieves a usable system capacity of 55%, has to be set against a greater circuit complexity.

8.6   VSAT REGULATION

The implementation of very small aperture terminal networks was delayed by the regulatory aspects of satellite systems and has only grown rapidly as the de-regulated environment in the USA, Europe and elsewhere has come about. Where de-regulation is not taking place there is still a move towards VSAT terminals, as the inherent advantages of the concept are realised and the pressure of the potential user begins to be felt. In countries, such as India, which have their own satellite systems, the technical direction of the VSAT will correspond to the frequency available on existing satellites. In these circumstances the supposed disadvantage of having a single regulatory authority in a country is not necessarily applicable as an overall planning operation can be carried out and if necessary new satellites can be planned.

The USA was the first area of the world to have a de-regulated environment, with the Federal Communications Commission (FCC) licensing the number of operators that were allowed to compete for business. In terms of VSATs the de-regulation perhaps went a stage further in that the FCC does not follow its usual practice of approving each terminal but only each type of earth station used in the overall network. This relaxation is in many aspects particular to the USA, which has a homogeneous population with businesses spread out across the country and using a single language. The European scene is different, not in that it is not deregulating because it is, but in that each country can have its own VSAT network, which may use TDMA, DAMA, FDMA/SCPC as its basic form of communication system bringing almost insuperable problems of interconnectivity across national borders. This position can be overcome by:

(a) The whole of Europe accepting a common VSAT network format – unlikely in the present climate.
(b) The continent of Europe could have a private satellite system that could be used for cross-border connections on a common basis. This may happen as dbs comes more into operation.

The other question raised by this is the manner in which VSATs are operated and their relationship with the existing INTELSAT network. INTELSAT has established its own INTELNET network which will cater for VSAT type network applications. In general the private VSAT network operator has to follow international and domestic rules. In international terms the regulations are defined by the International Telecommunications Union (ITU) which allocates the frequency spectrum that is available to the USA. The USA has, through the FCC, devised regulations that cover domestic and private networks operating on an international basis. The FCC began its regulation process in 1965, once INTELSAT began. The open skies policy began in 1972, a deregulating process that broke the monopoly of COMSAT and allowed any legal organisation to operate a satellite system. The FCC set technical parameters and satellite spacing. The process of deregulation was applied to the international market when in 1984 the USA set the rule governing the method by which the private operator could establish international satellite links outside INTELSAT and these guidelines allowed only transponder leasing. The private operator had to obtain authorisation from the foreign country and INTELSAT had to be considered to ensure non-interference with international communications via INTELSAT's worldwide network and the provision of public switched message services.

The other element of regulation has nothing to do with technical parameters but is environmental and to do with local community laws which restrict the placing of antennas on one's property. These restrictions are aesthetic, in that the size of the antenna is restricted to prevent visible eyesores being erected; on safety grounds in that the larger the antenna the greater the likelihood of it being blown down with consequent danger to life and limb; health aspects where microwave transmissions have to be controlled to avoid damage to individuals who may cross the transmission part of a badly-sited antenna; and on interference grounds where the earth station emissions may apply to both receive only and interactive VSATs.

8.7 TYPICAL VSAT SERVICES

8.7.1 The INTELNET service

The micro-terminal service set up by INTELSAT over the period 1985/86, offers two functional services: INTELSAT 1, which is a data distribution service utilising receive-only VSAT terminals and INTELNET 2, which is a data collection service. The VSAT terminals are interactive and can use both receive and transmit to a central hub, i.e. a Star network.

The space segments are available to the INTELSAT signatories on a leased basis. The leasing of transponders is arranged as for the normal domestic system and can be pre-emptible, i.e. the service can be withdrawn if INTEL-SAT need the capacity due to any emergency or non pre-emptible which guarantees the user a service regardless of circumstances.

The INTELNET service can be provided with the following character-istics:

(a) C Band – 4 GHz receive band/6 GHz transmit band;
(b) Ku Band – 11 GHz receive band/14 GHz transmit band;
(c) Ku Band – 12 GHz receive band/14 GHz transmit band;
(d) Satellite bandwidth in units of 5/9/18/72 MHz;
(e) Modulation can be:
  ● Binary phase shift keying – BPSK;
  ● Direct sequence binary phase shift keying – DSBPSK;
  ● Single channel per carrier BPSK – SCPC/BPSK;
(f) Forward error correction (FEC) at half rate;
(g) Data rate – 1.2 kbits – 512 kbits;
(h) Voice – 16 kbits and 32 kbits.

The INTELNET service does not provide mandatory definitions of the micro-terminal characteristics but sets the framework in which the network operator can use the leased satellite transponder capacity, with only those parameters that would adversely affect the satellite being set as a mandatory specification.

8.7.2 INTELNET terminals

The broad specification for INTELNET is defined as a Standard G in IESS – 601. This particular standard relates to earth stations that access INTEL-SAT satellites for international service not defined by any other INTELSAT standards. It is essentially a leased transponder service and it is non-specific in the following areas:

- Figure of merit;
- Modulation method;
- Transmit gain;
- Maximum eirp per carrier;
- Channel capacity.

The above parameters have to be set by the network user but must be approved by INTELSAT. The IESS, for the Standard G is a series of guidelines, which call for earth station flexibility in changing the carrier frequency, so as not to inconvenience INTELSAT when they need to change transponder allocations, as well as a number of mandatory design parameters which are listed in Table 8.1.

8.7.3 Typical terminal characteristics

In the data distribution system the up-link budget may be discounted when calculating total C/N of the overall system provided that the receive terminal antenna is small. This will result in the downlink controlling the C/N. The size of antenna depends to some extent upon the modulation methods used but is also limited by the interference limits set by the satellite system design.

BPSK terminal

Parameter	Value
Modulation	BPSK
FEC	$\frac{1}{2}$ rate
Data rate	64 kbits
Bandwidth	180 kHz
Bit error rate	10^{-7}
Eb/No	7 dB
C/N	3.2 dB
Antenna size	1.5 m
G/T (4 GHz)	12.0 dB/K
G/T (11 GHz)	18.7 dB/K

CDMA terminal

Parameter	Value
Modulation	DS/BPSK
FEC	None
Data rate	9.6 kbits
Bandwidth	5 MHz
C/N	− 12.9 dB
Antenna size	1.2 m
G/T (4 GHz)	11.0 dB/°K

It can be seen that the DS/BPSK system, while able to operate at much lower carrier to noise ratios, uses very much more bandwidth than a BPSK system with FEC and is therefore much more expensive to operate.

8.7.4 EUTELSAT terminals

The satellite distribution system used on EUTELSAT 1 for data services could not be described as a VSAT system in that it required an antenna size of 1.8 m

Table 8.1 INTELNET system parameters

Parameter	
Tx sidelobe	$G = 32{-}25 \log \theta$ $1° < \theta < 48°$ $- 10$ dBi $\theta > 48°$
Rx sidelobe	G as for the transmit sidelobe but not mandatory.
Antenna axial ratio	1.4 4/6 GHz IS4A 1.06 4/6 GHz IS5/5A/5A (IBS)/6 > 31.6 dB 11/14 GHz and 12/14 GHz The above specification can be less restrictive for antennas of less than 2.5 m diameter, in order to reduce feed costs.
Pointing capability	This should be sufficient to prevent outage during change of satellite. The accuracy of pointing must take account of satellite movement, eirp stability and axial ratio. The levels are defined elsewhere.

Frequency	C Band Rx 3.625–4.2 GHz
	      Tx 5.850–6.425 GHz
	Ku Band Rx 10.95–11.2/11.45–11.7 GHz
	      Tx 14.00–14.50 GHz
	K Band Rx 11.7–11.95/12.5–12.75 GHz
	      Tx 14.00–14.50 GHz
	The C Band frequencies are extended to allow operation on transponders 1 and 2 of INTELSAT 6 and will have an effect on feed design.

Transmission time delay	Satellite Max delay dTd (ms) (ns/s) 4A/5/5A/5A (IBS) 0.6 18.0 6 0.34 9.7 Delay is sum of up and down link delay.
Off-axis emission	At 6 GHz to CCIR Rec 524–2 At 14 GHz $D = 36 - 25 \log \theta$ $2.5 < \theta < 48°$ $- 9$ dBi $\theta > 48°$
Intermodulation	< 10 dBW in 4 kHz
eirp variation	> 15 dB
eirp stability	+ 1 dB/ − 1.5 dB

minimum for the remote terminals and an 8 m antenna for the hub station in a star network. With the EUTELSAT 2 series of satellites, with their higher eirp, the size of both remote and hub station antennas could be considerably reduced. The use of EUTELSAT as a satellite provider for a VSAT network operator would come under the rules and conditions laid down for a closed network, whose basic characteristics have already been described in chapter 7.

The original satellites tended to be power limited rather than bandwidth limited. This is reversed in the EUTELSAT 2 series of satellites which means that a VSAT system, using Ku Band, is much more suitable for use with access methods other than spread spectrum.

The terminals can operate in a simplex or duplex mode, with the use of antennas below the size of Standard 3. A typical link budget is shown in Table 8.2.

Table 8.2   EUTELSAT small terminal link budget

Parameter	Value
Rx	
Antenna gain 1.8 m	49 dBi
Pointing loss	0.3 dB
Feed loss	0.5 dB
System noise temperature	160 °K
G/T clear sky	22.5 dB/°K
Tx	
Antenna gain	50 dBi
Pointing loss	0.6 dB
HPA O/P to feed loss	0.4 dB
HPA output	10 W
eirp	59 dBW

Table 8.3   EUTELSAT hub station parameters

Parameter	Value
Antenna size	3.0 m
Remote antenna size	1.2 m
G/T	26.3 dB/°K
LNA noise temperature	120 °K
Maximum erip	48.6 dBW
HPA output power	7.5 W

The eirp in Table 8.2 takes no account of the HPA back-off required and has therefore to be reduced by 3.5 dB for a single carrier system. The hub station size can be very small as shown in Table 8.3.

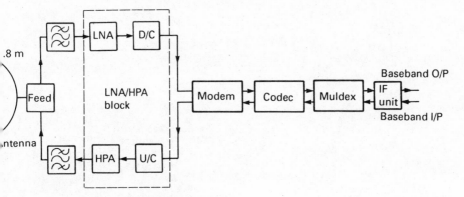

Fig. 8.3 Typical VSAT earth station configuration.

8.8 PRACTICAL TERMINAL DESIGN

8.8.1 Introduction

The VSAT earth station design follows the same format as any other earth station and a general form of the equipment configuration is shown in Fig. 8.3.

The system is divided into two blocks: an indoor unit that comprises the antenna, LNA or LNB and a downconverter; an indoor unit that comprises a modem, a codec, a muldex; and an interface unit. The antenna would most probably be an offset front-fed parabolic reflector and typically would use a 1.2 m or 1.8 m diameter reflector. The receive front end and transmit upconverter can be housed in a single unit mounted on the feed support arm to reduce the eirp and input noise temperature losses to a minimum. This is shown in Fig. 8.4, which also includes the filtering, orthogonal mode transducer (OMT) and polarisation adjustment.

The low noise block is limited in performance in two ways: its own basic noise ceiling due to its construction and semiconductor elements used; and the noise contributed to the input by the antenna itself. This restricts the practical reduction that can be obtained by the use of very low noise amplifiers because in poor weather conditions the antenna contributes more noise than the LNA and becomes the dominant element. The system designer must therefore take note of this in determining the trade-off between LNA noise performance and antenna size. There is a positive advantage in limiting the reduction in the earth station antenna size. The HPA output ranges between 1 and 3 W, using solid state components to give small size, high reliability and long life. The system G/T ranges between 17 and 21 dB/°K, depending upon the size of antenna and LNB.

The modem provides signal processing that is generally bi-phase PSK, though many suppliers are now providing the user with QPSK, which has an operational advantage in that it can provide more bits per Hz than BPSK.

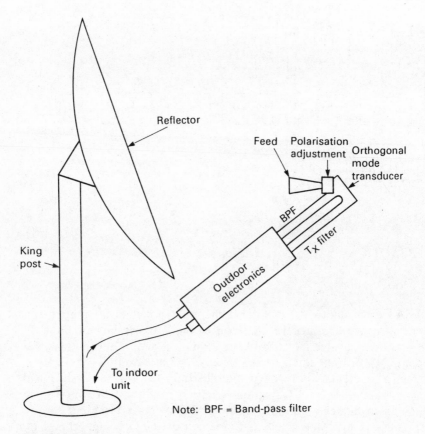

Fig. 8.4 VSAT outdoor system.

All systems use error correction in the form of convolutional coding, rather than some form of ARQ which would be ineficient in the amount of system throughput.

In the following paragraphs the technical details of the VSAT are taken from the system designed by COMSAT, which has a trade name of Starcom. In fact the terminal uses the expertise of two major communication companies: COMSAT who have a long history of involvement in the field of satellite communication systems and provided the expertise for the TDMA design and MELCO who have expertise in RF sub-systems and have designed the outdoor components, used in the VSAT terminal.

8.8.2 VSAT antenna

A typical antenna performance specification is shown in Table 8.4 for operation at Ku Band.

Table 8.4   Typical VSAT antenna performance

Parameter	Value
Diameter	1.2 m/1.8 m
Frequency bands	Rx 11.7–12.2 GHz
	Tx 14.0–14.5 GHz
Antenna gain Tx	43.0 (46.1) dBi at 14.5 GHz
Rx	41.5 (44.6) dBi at 12.2 GHz
Polarisation	Linear orthogonal
Sidelobe performance	To FCC spec for 2° spacing
OMT port VSWR	1.35:1

Table 8.5   Typical VSAT LNC performance

Parameter	Value
Input frequency range	11.7–12.2 GHz
LNC noise temperature	250 °K at 20°
LNC gain	74 dB
LNC bandwidth	500 MHz to 1 dB points

8.8.3   VSAT low noise converter

The LNC will probably employ GaAs FET technology to provide low noise performance, wide bandwidth and high gain with an IF signal output that can interface into the main modem block. A typical performance is shown in Table 8.5.

8.8.4   VSAT figure of merit

The figure of merit is detailed for both antenna sizes in Table 8.6.

Table 8.6   VSAT figure of merit for 1.2/1.8 m antennas

Parameter	Value
Figure of merit	19.2 dB/°K at 12.2 GHz for 1.8 m
	16.1 dB/°K at 12.2 GHz for 1.2 m
Noise temperature	250 °K
Pointing angle	> 20°
Transmission conditions	Clear sky

The figure of merit is the best case condition and for operational systems the rain conditions etc. must be taken into account.

8.8.5   VSAT high power amplifier performance

The HPA must not only provide a particular power output but it must provide stable gain, wide bandwidth and good gain slope performance to avoid intermodulation distortion. It is also necessary to restrict the noise/spurious outputs from the amplifier. The general performance of a typical system is shown in Table 8.7.

Table 8.7   Typical VSAT HPA performance

Parameter	Value
HPA output	2.2 W
HPA gain	48 dB
HPA bandwidth	500 MHz
Gain slope	0.3 dB per MHz
Noise output	− 70 dBm/4 kHz band
Antenna gain	43 dB at 14.5 GHz
eirp	46.1 dB
Transmit/receive isolation	80 dB

8.8.6   Signal processing systems

The signal processing is carried out in two forms: modulation of the data signals, at 56 kbits, on to the carrier signal, the modulation taking the form of BPSK and coding, which is a convolutional $\frac{1}{2}$ rate code.

8.8.7   Hub station

Introduction

The hub station design is a function of the overall system requirements, in terms of number of remote terminals in use and the amount of data being transmitted around the system. The hub earth station utilises an antenna size that predicates the use of tracking to keep the system on track of the satellite. The equipment is connected in a 1:1 redundancy configuration to ensure that failure does not interrupt customer traffic. The hub consists of:

(a) A 5.5 m–11 m antenna able to track the satellite;
(b) A 3.5 dB LNA with synthesised downconversion, that allows rapid modification to the receive system when more remote terminals are added to the overall system;
(c) An HPA giving an output of 5.5 W fed by a synthesised upconverter;
(d) A step-track tracking system;

(e) A signal processing system that caters for both 256 kbits continuous mode and 56 kbits burst mode operation;
(f) An LNA with a 2 dB noise figure to give the required G/T;
(g) Baseband processing units that carry out the modulation, demodulation and coding process;
(h) A computer system that controls all the functions required for the operation of a VSAT network.

The system can operate in various modes:

Pre-assigned
This is the least flexible in that as long as the connection is required it remains in force but can at any time be re-assigned as necessary, though it is often based on a fixed period of operation, say every 24 hours. Pre-assignment is achieved by allocating fixed slots in the TDMA frame and this is then unaffected by system call processing and slot reservation requests. It is expected that pre-assignment will be applied to large capacity customers who require the constant use of a channel, for those customers who wish to be sure that their communications link is not interrupted or for links such as political hot lines which, although having a small usage must remain as a permanent link. The pre-assignment is in its way very flexible in the services that it can provide, being capable of operating over a wide range of bit rates to carry voice, data and video services. In addition the pre-assignment can be set up between single earth stations or on a broadcast basis.

Dynamic network management
This allows the operator to dynamically re-configure the network as required by new or existing customers without interfering with the existing services. This facility necessitates a rapid response to customer demand and this can be achieved by arranging for VSAT reference stations to transmit a time marker to designate where a change is to be made prior to re-setting the circuit configuration, a system which also takes into account the satellite transmission medium delay.

Demand assignment mode
This can be of two forms, fully variable or variable destination, the difference being that the former requires a single burst per channel while the latter has multiple bursts, though it is usual to operate a fully variable DAMA system. This has the advantage of requiring less terrestrial interfaces and satellite channels.

The size of the network that can be supported by the hub station depends upon the size of the central node computer, which can be added to if modular architecture is used.

System requirements

If a TDMA system is used for a VSAT system then it will operate on a similar principle to that of INTELSAT systems etc., with the same problems and trade-offs being applied. The overall network will have a primary control station that is shadowed by a second identical terminal which is linked to the PCS and can take over from it in the event of PCS failure. The frame architecture is made up of a TDMA frame which carries a reference burst to time the network and a number of signalling queuing bursts (QBs), one for each terminal in the network to control the timing for each traffic terminal and traffic bursts.

The individual frames are built up into a multiframe structure which is variable in length because of the variable number of queuing bursts in the frame. The efficiency of the system requires only a small number of QBs, i.e. the network will be less efficient as the size of the network grows assuming that the burst bit rate remains the same. The frame efficiency can be increased by increasing the burst frequency. The frame efficiency can be derived from:

$$F_e = R_b \times t_f - B_s - \left(B_{cy} \times \frac{B_p}{R_b} \right) \tag{8.1}$$

where R_b = bit rate of burst

t_f = frame time

B_s = bit in signalling bursts

B_{cf} = total customer bits/frame

B_f = preamble bits

Satellite ranging

To ensure that the time slots are always in their correct position the effect of the satellite delay must be taken into account. Usually a number of terminals in the network are nominated as ranging terminals, chosen for their geographical position in the network to provide triangulation of the ranging system.

A ranging time interval is set as a multiple of a multiframe burst and the reference terminal receives from the ranging terminals its current value of distance to the satellite. With this information the central terminal computes the satellite's actual co-ordinates and this is re-broadcast from the control to the other terminals, which allows them to correct their acquisition system and be in synchronism with the control terminal.

The hub station control system can usually operate in an automatic redundant mode which means that it can set up the system, set up the communication links as required, accept new terminals into the network and, by use of additional modules, ensure that any failure is automatically overcome.

Chapter 9
Satellite Television

9.1 INTRODUCTION

The major use of satellite communications has always been for voice and data communications, with the development of domestic, regional and worldwide systems. Television transmissions were originally used to provide wide dissemination of special events on an irregular basis, probably due to the high cost of such a service which has the wide bandwidth requirements for transmission of a video signal. The use of satellites for TV transmission gained its biggest impetus in the USA with the coming of cable television, generating a need for a high quality TV picture that could be simultaneously connected to a large number of customers. The initial method for this was to use video recordings which, because of their reduced quality actually hindered the development of cable systems in the USA. The use of C Band satellites allowed a centrally generated programme to be transmitted to a number of cable head ends, or even private individuals who could afford the receiving equipment, which required large receiving antennas and high performance receiving systems. The second impetus to satellite television was the use of a new frequency band, Ku, operating in the 11–12 GHz region. This move to a higher frequency band, although it brought with it problems of polarisation and the need to design earth stations that could work at the higher frequency, was a crucial factor in the development of satellite TV, in the field of both receive-only videoconferencing and domestic reception.

The introduction of Ku Band operation has also produced a rapid development of antenna designs, low noise amplifier technology and low cost TV demodulators, which are now used, not only in TVRO applications but also in high quality TV transmissions, transportable systems, educational systems and monitoring systems. The numbers and types of application have increased considerably as performance has improved and size and cost of hardware have decreased. It is the domestic area of usage that has made the biggest impact and has provided the main commercial thrust of systems such as EUTELSAT, where satellite data services may struggle in competition with terrestrial services such as cable, microwave and optical fibre, but TV distribution from point to multipoint provides enormous revenue.

9.2 SATELLITE TV SERVICES AVAILABLE

The bandwidth requirements of a television signal mean that it is costly to use a satellite and a variety of services have evolved to provide a wide range of options for the satellite user and allow them to tailor their own services in the most economical manner.

9.2.1 Full time transponder usage

The major users of such a service are countries who require a nation-wide television service, without the need to actually own a satellite; countries who have very large areas to cover, which would require a very expensive terrestrial system to provide total coverage or countries that have a poor telecommunications structure and wish to install a system in as short a time as possible. At present these services are provided by INTELSAT who lease transponder time, on global or spot beams as required, both to individual countries, news organisations and commercial satellite television companies. This growth of satellite television has meant that INTELSAT now has a full-time, dedicated network that carries programming, mainly news, on a worldwide basis. Prior to this dedicated network services of INTELSAT were leased on a pre-emptible basis which allowed it to withdraw those services should the normal telephony transponders fail and the TV transponder need to be used as a back-up facility.

The normal bandwidth leased for television services is 36 MHz and on an INTELSAT global beam it provides 23.5–26.5 dBW of eirp from their later satellites. The second option available is to use a half transponder system providing 17.5 MHz bandwidth which allows the user to have a colour transmission with a number of voice circuits. This means that, from INTELSAT's point of view, they can offer a service of two TV transmissions in a full 36 MHz transponder, even though this results in the halving of the signal level available as the power is split between the two signals. In addition to the 3 dB power loss it is also essential that the signal power be backed off by a further 3 dB, to prevent interference between the two TV signals co-located in the same transponder. The growth of news services on a global basis has led to a further technical development which allows four sets of video and audio signals to be transmitted through a single 36 MHz transponder.

Other satellite systems, such as EUTELSAT, also provide full time leases of transponders, in their case to the European Broadcasting Union (EBU). In this case it is not just a matter of long term leasing of two transponders but the provision of very high quality TV services between EBU members.

9.2.2 Occasional transponder usage

There is a considerable need for occasional use of transponders for TV transmissions, particularly for news feeds, on an intercontinental basis. It is a service that broadcasters use to cover special events, so that the satellite provider needs to be able to allocate bandwidth at short notice. It also means that the satellite transmission levels must be sufficient to provide a high quality service (S/N > 50 dB) within the satellite footprint.

9.2.3 Subscription transponder leasing

A half-way house between full time and occasional leasing is the planned part-time use of transponders, usually with a minimum rental time. This service is used for special events transmission, intra-company videoconferencing and general TV transmissions.

9.3 TRANSMISSION CHARACTERISTICS

The range of possibilities for satellite TV transmission is very wide, from systems using 3 kW amplifiers for transmission and high quality, low noise threshold extension receiving systems to a viewable quality system used for temporary transportable links designed for crisis situations. This is exemplified in the EUTELSAT system, which for EBU transmissions operates with video S/N ratios of 53 dB for 99% of the worst month of the year, implying a 15.5 dB C/N, well above threshold, but for transportable systems it allows a S/N of 50 dB. This disparity of standards according to usage, is now common and each system should be designed according to the minimum requirements.

9.3.1 Television standards

One of the reasons for the slow growth of international television is the variety of television standards that apply to the television picture. There are currently three standards; the first standard was set by the USA in 1940 when the National Television Standards Committee (NTSC) was formed to approve a system and recommended a 525-line system at 30 frames per second, a system that was made compatible with colour transmissions in the 1950s. The second standard, set in Europe by West Germany, attempted to resolve the problems of phase errors by producing the phase alteration by line (PAL) system that cancelled out phase errors in a colour signal by alternate line phase inversion. However due to the inability of the various countries to agree on a common system the French produced a third standard, using frequency, not amplitude modulation, called sequence with memory (SECAM). This multiplicity of standards has been an international nightmare for colour TV transmissions, though it did in fact generate a new industry manufacturing standards converters and led to manufacturers, especially in Europe, producing multi-standard television sets. It is interesting to note that for satellite transmissions a similar problem arose with regard to the special techniques being devised for transmission of video signals. These methods were considered in chapter 5.

9.4 SATELLITE TELEVISION SYSTEMS

It is important to distinguish between the distribution of television programming by means of satellite and dbs in order to appreciate the different equipment requirements for each system.

The fundamental difference between the two forms of system is that the dbs system satellite footprint is shaped to cover only a specific area, such as a single country and to provide a high level of illumination at the earth, while the satellite TV distribution system is not defined so specifically and has a lower level of illumination, necessitating larger antenna systems.

9.4.1 Direct broadcasting by satellite (dbs)

Introduction

The allocation of frequencies and number of channels to each country is performed by the ITU organisation with the actual process of allocation being carried out at a series of World Administrative Radio Conferences (WARC).

The ITU has divided the world into three regions for regulatory purposes. The three regions were allocated a frequency band, number of channels, polarisations, receive levels and antenna size. The 1979 conference reviewed the technology available and the cost of such technology; at that time the technology of LNCs in particular appeared to indicate an antenna size of 90 cm. That estimate was incorrect and it is now possible to envisage 60 cm antennas, 100 °K LNCs and multi-channel receivers at a cost below the level of a colour television set. It is important to realise that for dbs, which is aimed at a mass market, the technology used is not the important parameter except in so far as it contributes to the major element, low cost, without which the consumer will not take up the product offered, i.e. television programming. Indeed the problems of dbs are not in the technology but in the marketing of the services, especially when they are in competition with the terrestrial services, cable systems and satellite TV distribution.

The first WARC meetings took place in 1977 with the express purpose of defining the regulations that would apply to satellite broadcasting. In detail the three regions are:

- Region 1 – Europe, Middle East, Africa, USSR and Mongolia;
- Region 2 – North and South America and Greenland;
- Region 3 – India, Iran, South East Asia, Australasia, Japan, China, Pacific.

The 1977 conference dealt with Regions 1 and 3 and produced a plan that allocated the frequencies to each region as shown in Table 9.1.

Table 9.1 WARC regional dbs frequency allocations

Region	Frequency GHz
1	11.7–12.5
2	11.7–12.1
3	11.7–12.2

The performance specification is also set by WARC and the basic system parameters are shown in Table 9.2.

Table 9.2 WARC Region 1 dbs Rx design parameters

Parameter	Value	
G/T	6 dB/°K	
S/N	33 dB	
LNA noise figure	3 dB	
C/N	14 dB	
Received PFD Reg 1	− 100 dBW/m sq	BC
	− 103 dBW/m sq	BE
Reg 2	− 102 dBW/m sq	BC
	− 105 dBW/m sq	BE
Reg 3	As Region 1	
Co-channel interference	30 dB	
Adjacent channel interference	14 dB	

The downlink parameters were set by WARC in 1977 but did not initially set the two uplink frequencies. The WARC 1979 conference allocated two uplink frequency bands; 10.7–11.7 GHz and 17.3–18.1 GHz. In practice this meant that the latter band was used because the lower frequency band clashes with the 11 GHz terrestrial services and the Ku Band fixed satellite service.

The parameters for Region 2 were not decided until a Regional Administrative Radio Conference (RARC) in 1983. The parameters set for Region 2 are given in Table 9.3.

Table 9.3 **WARC Region 2 dbs Rx design parameters**

Parameter	Value
Receive G/T	10 dB/°K
S/N	33 dB
LNA noise figure	3 dB
C/N	14 dB
Rx PFD	− 107 dBW
Polarisation	Circular
Channel spacing	14.58 MHz

Region 2 was allowed to have satellites on an orbital arc from 31° W to 175° W with 48 satellite positions providing 32 channels per position. The transmissions are made with both left and right circular polarisations. The channel bandwidth is set at 24 MHz with guard bands of 12 MHz between channels.

Direct broadcasting in the USA

In the USA the dbs systems are not yet fully operational but the first proposals made, by eight separate applicants, are shown in Table 9.4. It should be noted that the majority of these applications propose the use of high power TWT amplifiers on the satellite and it is this element that has delayed the implementation of these systems. This is because it is a high-cost solution and therefore difficult to finance, particularly as the USA already has a mature satellite television distribution system, linked into a large number of cable networks.

The systems shown in Table 9.4 not only intend to transmit standard TV programming but also a variety of experiments associated with high definition television (HDTV). The technical parameters of each system highlight the problems of obtaining a uniformity of standards that have bedevilled the television industry since its inception, as well as the fact that in certain cases the systems are almost not true dbs having antennas on the limit or just above the WARC recommended size. The element that has changed is the cost and performance of the low noise front end systems and these are now available at low cost and with a 150 °K noise temperature. This would give a 3 dB improvement in performance and allow either the satellite power or the receiver antenna diameter to be halved.

Table 9.4 Proposed USA dbs systems

Operator	TWT (W)	Channel	BW (MHz)	Polarisation	eirp (dBW)	SATG/T (dB/°K)	E/S G/T (dB/°K)	C/N (dB)	Antenna diameter (m)	°K
USSB	230	3	16	C	57	7.7 dB	8.8	15.4	0.75	550
Sat TV	215	3	16	C	57	7.7	9.4	15.9	0.75	527
RCA	230	4	24	L	–	– 3.6	10	15.6	0.6	500
Western Union	100	4	16	L	55.5	7.0	10.4	14.0	0.9	550
DBS Corp	200	6	22.5	C	56	5.0	12	15.7	0.9	420
Vid/Sat	200	2	36	C	56	– 3.6	10.2	13	0.9	552
CBS	400	3	27	C	60	8.2	12	19.8	1.0	–
Gr/Scan	300	2	18	C	53.7	8.1	10	15.2	0.9	440

Direct broadcasting in Europe

Europe was further ahead than the USA in implementing true dbs system as opposed to satellite television relay systems, though the problems of implementation are identical in that the main questions to be answered are financial and to do with programming rather than technical difficulties. There are a number of systems already in place, two major ones are the ASTRA satellite system and the EUTELSAT system.

ASTRA satellite system

The ASTRA system is centred in Luxembourg with its satellite positioned at 19.2° E. The satellite antennas are designed to give a broad European coverage, with transmission on both horizontal and vertical polarisations. The contours of satellite eirp for each of the four modes are shown in Fig. 9.1, this being done in order to give the individual user between eight and 12 channels, received on a 60 cm antenna with the remaining 4–8 channels requiring a larger antenna for better reception.

The various transmission patterns reflect the need to obtain as much market as possible to have a viable service covering the majority. The satellite has 16 transponders, using a 45 W TWTA and having a bandwidth of 27 MHz. The downlink frequencies occupy the frequency band between 11.20 GHz and 11.45 GHz. In Table 9.5 the link budget parameters are listed for a 65 cm antenna and assumed antenna efficiency and LNC characteristics.

Table 9.5   ASTRA dbs link budget

Parameters	Value
Antenna diameter (m)	0.65
Antenna efficiency (%)	65
eirp (dBW)	52
System noise temperature (°K)	160
Receiver bandwidth (MHz)	26
Transmission attenuation (dB)	205.3
System losses (dB)	0.6
Carrier to noise C/No (dB)	13.0
Modulation/pre-emph/weighting gain (dB)	28.7
Video bandwidth (MHz)	5.0
Video deviation (MHz/p–p)	13.5
Energy dispersal (MHz/p–p)	2.5
Weighted S/No (dB)	43.7

The parameters in Table 9.5 assume PAL transmissions, though a MAC transmission could improve performance, allowing a smaller antenna to be used for reception. The relationship between antenna diameter and satellite eirp is shown in Fig. 9.2.

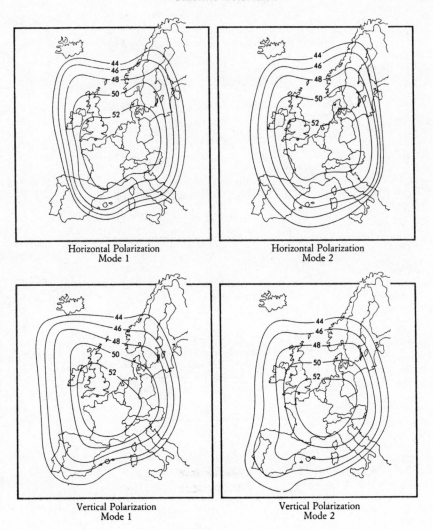

Horizontal Polarization
Mode 1

Horizontal Polarization
Mode 2

Vertical Polarization
Mode 1

Vertical Polarization
Mode 2

Fig. 9.1 ASTRA satellite footprints.

The quality of the signal is defined by the weighted signal to noise ratio and this ratio has been related to the subjective picture quality in CCIR Report 959-1, as detailed in Table 9.6.

The table shows that for a typical ASTRA reception system picture grade will be good/excellent, assuming clear sky conditions. The picture quality will degrade due to rain attenuation and this can be allowed for by reference to Table 9.7.

The figures in Table 9.7 can be used to design the receiving system to the particular grade of service required, whether individual or communal, though it should be recognised that for communal installations, either cable

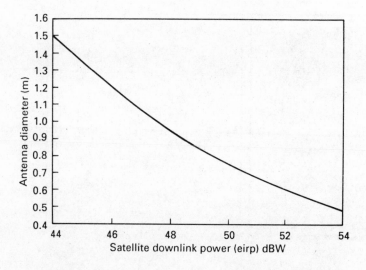

Fig. 9.2 ASTRA satellite eirp/required antenna size.

Table 9.6 **Relationship of picture quality to baseband S/N ratio**

Picture grade	Quality	S/N (dB)
1	Bad	29.3
2	Poor	33.6
3	Fair	38.0
4	Good	42.3
5	Excellent	46.6

Table 9.7 **Rain attenuation effect on S/N ratio**

%/year	S/N reduction (dB)
99.0	0.8
99.2	1.05
99.4	1.30
99.6	1.55
99.8	1.95
99.9	2.45
99.95	3.50

or satellite master television (SMATV), the grade of service should provide a good or superior quality picture for 99.95% of the time, i.e. a 0.8 m antenna would be required at the centre of the footprint. A design guide to antenna size, for various values of eirp is shown in Table 9.8.

Table 9.8 Availability of good quality picture/antenna diameter and eirp

%/year	Antenna diameter (m)		
	48 dBW	50 dBW	52 dBW
99.0	0.89	–	–
99.4	0.95	0.76	0.6
99.8	1.04	0.82	0.6
99.9	1.11	0.89	0.66
99.95	1.30	0.95	0.80

Note: The above figures take into account rainfall effects.

EUTELSAT system

The satellite transmissions of EUTELSAT are based on a long experience, in satellite terms, of operating such services. The first TV transmissions were not intended to be mainly for individual reception, though in practice this did occur. The second generation of EUTELSAT satellites is intended to build on the unexpected success of the TV distribution service built up on the EUTELSAT 1 series of satellites. The satellites will provide a high gain beam coverage, as shown in Fig. 9.3, using 50 W TWTAs to give footprints very similar to the ASTRA system.

The basic parameters for the EUTELSAT system are shown in Table 9.9

Table 9.9 EUTELSAT 2 Rx system requirements

Antenna diameter (m)	C/N ratio (dB)		
Rx noise temp °K	100	150	200
0.6	9.9	8.7	7.7
0.75	11.8	10.6	9.7
0.90	13.4	12.2	11.3
1.20	15.9	14.7	13.7

which shows the antenna size, receive noise temperature and C/N for the following conditions:

- High gain contour 50 dB;
- Satellite at 12° E;

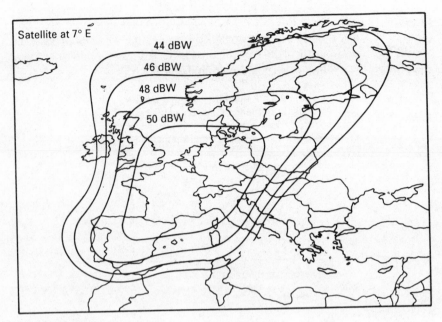

Fig. 9.3 EUTELSAT 2 high gain coverage.

- Receiver location gives 30° elevation angle;
- Antenna efficiency 65%.

The effect of geographical location on performance is not as extreme as one might expect, being only 0.5 dB over the whole of the 50 dBW footprint. The EUTELSAT 2 system has to cater for three television standards, PAL, SECAM and MAC, each requiring different bandwidths and receiving equipment. The PAL or SECAM system design will have a link budget as shown in Table 9.10.

The system design aim is to achieve a good or better than good quality. There must therefore be a system signal to noise value of greater than 42.3 dB; for the PAL/SECAM transmission can use either a 27 MHz or 36 MHz bandwidth though the emphasis of the system design will change depending upon the bandwidth used. If a 27 MHz bandwidth is used then the carrier to noise must be relatively high in order to achieve the required S/N of 43 dB. If a 36 MHz bandwidth is used then the effective antenna diameter must be large enough to ensure that the system remains above the FM threshold, which will mean that the baseband S/N will be well above 43 dB and will give a performance that is almost excellent quality. The MAC system will give a better performance than PAL or SECAM in terms of picture quality, but will have a problem at low carrier to noise ratios because there will be data in the signal and therefore the C/N value must be above 10 dB.

Table 9.10   EUTELSAT 2 link budgets for 27/36 MHz bandwidths

Bandwidth (MHz)	27	36
Parameter	Value	
Satellite eirp (dBW)	50	50
Path loss (99.5%/yr) (dB)	207	207
Antenna diameter (m)	0.75	0.65
Antenna gain (60% efficient)	37.1	35.9
LNC noise temperature (°K)	110	110
Antenna diameter noise temperature (°K)	50	50
Rain noise temperature (°K)	90	90
Pointing loss (dB)	0.3	0.3
W/G loss (dB)	0.3	0.3
TVRO G/T (dB/°K)	14.8	13.5
Noise BW (dB/Hz)	74.3	75.3
C/No (dB)	10.6	8.0
Video B/W (MHz)	5	5
Video deviation (MHz/volt)	16.0	25.0
S/No (dB)	43.0	45.5
FM modulation gain (dB)	19.2	24.3
Pre-emphasis gain (dB)	2.0	2.0
CCIR weighting factor (dB)	11.2	11.2

9.4.2   Factors affecting dbs systems

Introduction

The fundamental requirement of a dbs system is that the television service should be available twenty-four hours a day and should not require complicated reception systems, either in terms of the receiver technology, antenna design or pointing requirements. The antenna, must be of small physical size, lending itself to fixed parabolic or flat plate designs because of the wide antenna beamwidth, thus facilitating mounting on houses rather than an installation at ground level, which must cause minimum obstruction. Although CCIR has considered the use of sub-synchronous orbits it is improbable that they would be used in a commercial environment. This requires not only the maximum programming time – to make optimum use of the capital involved in the satellite and its launch – but also inexpensive ground installation whether on an individual or community cable level.

Frequency of operation

The introduction of dbs and satellite services comes at a time when the use of Ku Band is increasing in order to alleviate the congestion being experienced in the 4–6 GHz band. It can be seen from the WARC allocations that it is intended that the majority of dbs services are carried on Ku Band satellites. The question of the effects of the transmission medium on the satellite signal has been considered in chapter 2 where the effects of rain etc. were detailed. There have been extensive measurements made, using the OTS satellite operating at 11.5 GHz, to provide design information for the European region thus allowing the system designer to set the level of availability. The attenuation at 11.5 GHz is given in Table 9.11.

Table 9.11 Ku Band availability/attenuation characteristics

% time (minutes)	Availability (%)	Medial level attenuation (dB)
2160	95.0	0.6
432	99.0	1.1
130	99.7	1.8
43	99.9	3.3
13	99.97	7.3

The level of availability in the worst month is set as a system design parameter and is thus under the control of the designer. It will also be necessary to take into account the effect of frequency changes and this can be calculated from equation 9.1.

$$L_{db} = L_i[1 + 0.2(f_a - 11.5)] \tag{9.1}$$

where L = loss at any frequency > 11.0–14.5 GHz

L_i = loss at 11.5 GHz

f_a = actual frequency of operation

For Region 3 other measurements have been made which show that for moderate climates the attenuation is approximately 1.5 dB higher and for equatorial regions the attenuation must be increased by approximately 1.0 dB compared to Europe.

Polarisation

The effect of depolarisation has already been referred to in chapter 2, it is sufficient to say that cross-polarisation is a function of frequency, polarisation tilt, rain distribution and direction of pointing. There are a number of factors that have to be considered when deciding which polarisation to use for dbs systems, which are accentuated by the fact that it is a system with very small receiving antennas and therefore more sensitive to adverse conditions. The larger antenna systems, which operate with a linearly polarised signal, usually have some form of polarisation control to minimise the degradation in system performance. This adds both to the cost of the antenna and the cost of installation of the system, which must have a detrimental effect on the take-up of any dbs service.

The antenna shape becomes important in small antenna design and research has concentrated on the development of highly efficient antennas of small physical size using elliptical reflectors. The use of these makes the design of the cross-polar performance more difficult than traditional circular reflectors, as the plane of polarisation varies with elevation angle and does not always correspond to the major axis.

Satellite TV antennas often have a requirement to be used with two satellites and therefore must be capable of re-alignment to a new physical position. This process is often not a simple one, especially when linear polarisation is involved due to the need to reset the antenna polarisation that optimises the antenna performance for the new polarisation.

The effects of rain were considered in chapter 2 and one effect is depolarisation, which is worse for circularly polarised systems than linear ones, in fact there can be up to 15 dB difference between the two. The effect of depolarisation has been measured under a variety of conditions and can be calculated from equation 9.2.

$$D_c = X - 20 \log A \qquad (9.2)$$

where A = attenuation in dB at 11 GHz

$X = 30 \log f - 40 \log (\operatorname{cosec} e) - 20 \log (\sin 2t)$

f = frequency in MHz

e = elevation angle

t = polarisation tilt angle with respect to horizontal

This formula applies to circular polarisation and in practice is a much more random phenomenon than that derived from equation 9.2. For design purposes the designer has to decide what availability is required and choose a particular worst case condition. For linearly polarised signals the cross-polar discrimination is given by equation 9.3.

$$D_1 = 41 - 20 \log A \text{ dB} \qquad (9.3)$$

This is a pessimistic figure compared to actual measurements. The degree to which cross-polar/attenuation relationship applies is a function of elevation angle.

Interference

The use of dbs brings with it a far greater possibility of interference with terrestrial systems or between two adjacent satellite transmissions due to the higher satellite power levels involved in illuminating a restricted portion of the earth's surface. This problem has led to close consideration of both the satellite position for specific transmissions and the characteristics of the receiver on the ground. The overall system designer not only has to take into account the receiver characteristics but must also decide what interference is allowable between adjacent satellites.

In the WARC allocations the general principle used is that the satellite transmits to a number of different countries with an allocation of channels per country. In this instance it is likely that the same polarisation would be used for each of the transmissions and in interference terms the type of polarisation used can have a considerable bearing in the performance. Circular polarisation will give a 3 dB reduction in interference compared to linearly polarised terrestrial services. The same factors apply to dbs as any other system when considering the satellite and channel spacing for minimum interference in that orbital position accuracy, propagation effects, characteristics of transmit and receive systems must all be taken into account. The allocation of specific channels to each country also becomes critical in that adjacent countries should not be allocated adjacent channels. This last requirement also places a constraint on the receiver tuning range if it is to receive all of the allotted channels. An improvement in performance is possible if orthogonal polarisations are used, both in the spacing of the dbs satellite and the reduction of terrestrial interference between satellite transmissions.

System considerations

Any transmission system is usually a compromise between the power available to transmit the signal and the bandwidth available to the frequency of transmission. The decisions made in this respect are not simply a matter of availability but also the system cost.

The power amplifiers in the satellite can use up to 240 W per channel and, although it would minimise the design and cost of the ground receiver, it would be very expensive in satellite terms. The other extreme is to use satellite amplifiers as low as 60 W, which certainly reduce the cost per transponder but could not then be called a true dbs system. It would require a ground receiving antenna that would be twice the size of a true dbs system and therefore be a satellite distribution system more suited to cable distribution than direct to the consumer. The compromise position is the use of 100 W transponders, which would allow the use of 75 cm antennas at the beam centre, as used in the Japanese dbs system using the Yuri 2a series of satellites. In practice there is no one solution to the problem of how to balance the cost of the satellite, where the major start up costs are to be found, and the need to have simple, low cost ground receivers which are a necessity if the service is to be taken by the individual consumer. The decision has to be taken with the market place in mind for if there are no users of the service and therefore no revenue then the highest of technology will be of no use.

The other element to be considered is bandwidth required, a very valuable commodity. Now high definition television is being defined as a worldwide standard it is even more important that it is used efficiently. For dbs the preferred method of modulation is frequency modulation, with a number of sound channels being transmitted so that stereophonic sound can be broadcast. The receive system must be designed so that the composite TV signal is recovered without loss of information. In practice the occupied bandwidth is virtually infinite for FM modulation. What is necessary is to be able to quantify the signal degradation due to bandwidth limitation. The design is a compromise, the system signal/noise performance is a function of the IF bandwidth but the improvement in S/N performance, due to FM deviation increase, requires a larger bandwidth. The system S/N performance is also affected by truncation noise, a parameter that is affected by two circuit elements which cause variations in the carrier level. If the signal is pre-emphasised, as the steepness of the bandpass filter skirt slope increases the transient carrier reduction during transient signal conditions increases thus reducing the C/No. If this is near to threshold it may cause serious distortion. If the pre-emphasis characteristic is altered then the C/N varies for the same transient conditions. The threshold of truncation noise for a particular bandwidth and at a specific carrier to noise are shown in Fig. 9.4.

The bandwidth required, as defined by Carson's Bandwidth Rule, is 27 MHz for a 625-line TV system with a 14 MHz/V p–p frequency deviation. This particular set of parameters gives a good performance with low differential gain and phase distortion. The sound signal performance will have to be as good as terrestrial systems and add minimally to the bandwidth required. There are a variety of methods open to the system designer when considering the transmission of sound channels, they are all various forms of sub-carriers multiplexed together with the video signal and modulated on to the RF carrier. The quality aimed for is the same as that obtained in any analogue FM system, that is:

Audio bandwidth	15 kHz
Signal/noise	50 dB
Pre-emphasis	50 microseconds

The above characteristics can be achieved for a 625-line system, with two multiplexed sound channels, by use of a 27 MHz bandwidth. This band-

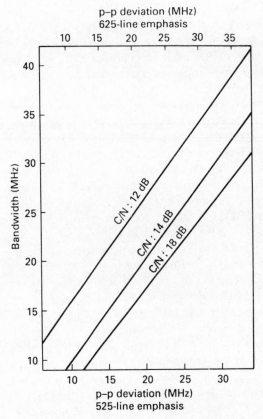

Fig. 9.4 Television bandwidth required/p–p deviation.

width would not only take account of the signal bandwidth required by Carson's rule but also allow for the variations in the frequency of the transmitter, while still providing sufficient bandwidth between channels to provide adequate adjacent channel interference protection. At the stated bandwidth it is possible to obtain a S/N ratio of 45 dB, at a point 4 dB above threshold.

It has been pointed out that the overall system design is a trade-off between the satellite complexity and earth terminal costs, to do this the designer must know the following parameters:

● Satellite eirp in dBW;
● Variations of signal level over the reception area in dB;
● Free space loss in dB;
● Attenuation margin due to rain dB;
● Figure of merit (G/T) for the dbs receiver in dB/°K;
● Receiver threshold point in dB;

- Losses due to depolarisation, antenna misalignment etc. in dB;
- Weighting and pre-emphasis.

With the above information and assuming a minimum luminance to weighted rms noise it is possible to design a dbs receiver. The overall system performance is defined by equation 9.4.

$$\frac{C}{N} = \frac{S}{N} - F_1 - F_2 \tag{9.4}$$

where $\dfrac{C}{N}$ = pre-detection carrier/noise ratio

$\dfrac{S}{N}$ = p–p luminance amplitude/weighted rms noise

F_1 = de-emphasis and weighting improvement

      For 625 line PAL $f_{vm} = 5\ MHz$ $F_1 = 13.2\ \text{dB}$
      For 625 line SECAM $f_{vm} = 6\ \text{MHz}$ $F_1 = 14.3\ \text{dB}$

$$F_2 = 10 \log \left[\frac{3B}{2} \times f_{vm} \times \left(\frac{B}{f_{vm}} - 2 \right)^2 \right]$$

      where B = peak to peak deviation (Dp–p) + $2f_{vm}$
                f_{vm} = maximum video frequency

To calculate the required figure of merit equation 9.5 must be used.

$$\frac{G}{T} = \frac{C}{N} - \text{eirp}_{\text{SAT}} + L + 10 \log(D_{pp} + 2f_{vm}) + 10 \log(228.6) \tag{9.5}$$

where L = free space attenuation dB
         = $92.44 + 20 \log f_t + 20 \log d_s$
    d_s = distance to satellite
        = $[(43.781 - \cos^2 e) - \sin e] \times R$ km
    e = elevation angle to satellite
    R = mean radius of the earth (6367 km)
    f_t = carrier frequency in GHz

The satellite eirp is defined in dBW and the actual signal at the earth's surface can be defined in two ways: in dBW/m² in a reference bandwidth, which is usually set at 1 Hz, or as a direct power. In general, information is readily available on satellite footprints for the majority of geostationary satellites now in orbit. The footprints reflect the satellite output power when the satellite transponders are operating at saturation although for many video transmissions the transponder does not operate at saturation but is backed-off from full power.

Sound sub-carrier systems

The basic FM system uses a number of sound channels, multiplexed together with the video signal before being modulated on to an RF carrier. There are a number of alternatives to this arrangement, which uses a digitally modulated sub-carrier. There have been a number of studies carried out looking at a variety of different modulation methods such as PCM, PSK and MSK. Apart from the modulation methods there are three basic ways in which the sound can be multiplexed with the video carrier: analogue FDM, digitally modulated signals in FDM mode, and digitally modulated signals in TDM mode. The digital methods require more band-width than FM systems. In addition it adds to the complexity of the dbs receiver, though it has the advantage that sound sub-carrier demodulation can provide high quality audio channels, especially if a QPSK system is used. This could be achieved by a VLSI device, one that would be low cost if made in sufficient quantities.

Basic multiplexing can be in two forms, continuous or packet. The continuous method utilises a frame system that contains a fixed number of bits, each of which has a pre-determined function, such as frame alignment. This method has the advantage that it allows the use of a simple receiver, has a low bit overhead and thus has a short synchronisation time. The packet multiplexing system uses blocks of information, a header block which synchronises the receiver and identifies the following block which contains the sound information. These systems suffer from the usual problems associated with any digital system, in that the synchronisation must be accomplished in a certain time and held in that state. This ability to remain in synchronisation is a function of signal bit error rate. The choice between these two systems is difficult and it will probably resolve itself to the simplest and most cost-effective receiver.

Receiver characteristics of dbs

The original WARC regulations envisaged a system which would provide a satellite with transponders operating in the 11.7–12.2 GHz range and providing a power output of 240 W. The LNA technology was assumed to give a 280 °K or 3 dB noise figure. The antenna size was set at 90 cm, it being assumed that it would be of parabolic form.

The technology has moved on much more rapidly than envisaged at the time of the WARC recommendations and therefore the designer has much more freedom to reduce satellite power or simplify receiver design. The basic dbs receiver system comprises the following major elements:

- Antenna;
- Low noise converter block (LNB);
- Indoor receiver;
- Interconnections.

These elements can be configured in a number of ways to achieve the best performance in individual circumstances.

Antenna systems

The antenna size can in practice be less than the 90 cm for dbs reception, depending upon satellite power, and is realised in three basic forms:

Centre-fed parabolic
The centre-fed parabolic can be constructed in various ways mainly related to the ratio of focus distance and parabola diameter, which ideally should be 0.375 to give the most efficient illumination. The system should therefore have a low F/D which will make the antenna less subject to interference and have lower noise temperature. The disadvantage of this system is that it suffers from loss of efficiency due to signal blockage and scattering of the signal by the feed support system.

Offset-fed antenna
The offset antenna reflector is shallower than the front-fed, as it is a section of a large prime focus parabola. This form of construction is more efficient than the front-fed system because it has no support struts. In addition it is easier to connect the LNB, which itself is more easily protected from the elements and overall has a lower cost due to its greater simplicity. The method of manufacture is simpler, being either a spinning or pressing of steel, or aluminium or fibreglass moulding. Once the basic reflector has been made then it is dip plastic coated to protect it from the elements. There is another environmental consideration that must be made and this is snow retention in the parabola and here again the offset parabola with its low F/D has better performance than a high F/D antenna. If the problem is resolved by the use of a radome then the offset system is again superior having no feed support system to impede the structure, though a radome is not a cost-effective solution to the problem.

Flat plate antennas
These are perhaps the ideal solution for a dbs antenna for the following reasons:

- Simple to mount;
- Low wind resistance;

- No feed blockage;
- Built-in LNB;
- Able to handle all forms of polarisation;
- Low environmental pollution.

A single polarisation receptor, using a 350 mm square flat plate gives 31 dBi, equivalent to a 0.4 m parabolic antenna. It should be noted that the same flat plate area applied to a parabolic antenna would give a surface area of 12.25 m², i.e. a parabolic reflector of 4 m, which would give a parabolic gain of 51 dB. The most difficult problem is the cost of the system which is not a function of the material used, but the processing involved in the construction of the antenna, which has four separate layers of material, each having a different design function. The detailed specification of a flat plate antenna is given in chapter 4.

One area of interesting research being carried out by a company known as *Mawzones*, is into the use of Fresnel rings as a means of focusing the signal from the satellite. The basic principle is based on the use of concentric circles of absorbent material, as shown in Fig. 9.5.

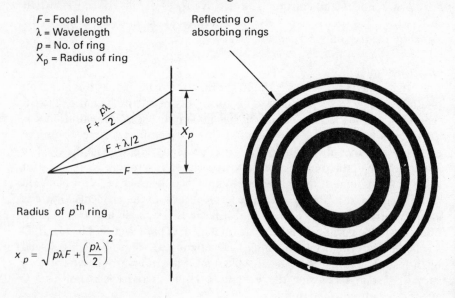

F = Focal length
λ = Wavelength
p = No. of ring
X_p = Radius of ring

Reflecting or absorbing rings

$F + \dfrac{p\lambda}{2}$

$F + \lambda/2$

X_p

F

Radius of pth ring

$$x_p = \sqrt{p\lambda F + \left(\frac{p\lambda}{2}\right)^2}$$

Basic zone plate lens

Fig. 9.5 Signal focusing by concentric circular zone plate.

In practice the antenna is made up of two plates, the first a Fresnel zone plate which acts as a focusing plate with the circle diameters as defined in Fig. 9.6. The second plate is a metalised sheet which has the same focus point as the Fresnel plate and therefore increases the gain of the overall antenna.

The linear form of antenna is satisfactory for signals that are perpendicular to the plate antenna.

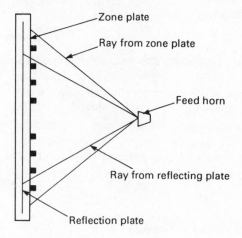

Fig. 9.6 Reflective zone plate focusing.

The antenna can be made more flexible by the use of an elliptical set of Fresnel rings, offset from the central axis, which will give signal focusing for signals that are not perpendicular to the antenna. The resulting offset antenna is shown in Fig. 9.7 together with the delineation of the signal paths.

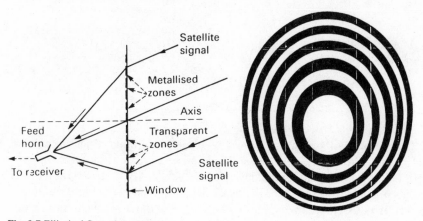

Fig. 9.7 Elliptical flate plate antenna.

The above antenna can be very flexible in its form, involving either a simple flat plate that can be wall or window mounted, or a mirror form that can be mounted on any flat surface that can see the satellite. In addition a matrix system can be mounted on any flat roof but will be considerably larger in size thus giving a much higher gain. These advantages have to be set alongside the ease of transportation and installation to provide a very competitive dbs antenna, which will not be subject to the planning difficulties that other parabolic antennas may have.

Low noise converter blocks
The development of low noise front end systems has been extremely rapid, pressured as it was by the need for low-cost, high-performance units for satellite distribution systems. It is one of the most important elements in any receiving system whether dbs or otherwise. The device should not only present as low as possible noise temperature, but also have a reasonable value of gain to ensure that the receiver noise figure gives only minimal reflection into the front end noise temperature. The noise figure is quoted as either a noise figure or a temperature and in Table 9.12 the correspondence between the two is shown.

Table 9.12   Noise temperature/noise figure equivalents

Noise temperature (°K)	Noise figure (dB)
20	0.22
40	0.56
60	0.82
80	1.06
100	1.29
120	1.51
140	1.72
160	1.91
200	2.28
250	2.71
300	3.09

The majority of systems now use low noise block units which take in the SHF signal at either C Band or Ku Band and provide an output frequency of 950–1750 MHz. In general the device gain is 45–50 dB, with variation of the gain being held to within 1 dB in any 27 MHz segment of the input frequency range. It is essential to recognise that the Ku Band frequencies are not uniform throughout the world and the LNB must be correct for its geographical area of usage.

The performance of LNBs were originally based on GaAs FET technology which has been refined to allow devices to operate in Ku Band with noise temperature of better than 120 °K having a physical size of 180 mm × 60 mm × 60 mm at a cost of less than £100 (1988 prices).

Receivers
The essential characteristics of dbs receivers are:

- Selectable multi-channel operation;
- Inexpensive;
- Wide input frequency range 950–1750 MHz;
- Automatic tuning;
- Input level -60 to -20 dBm;
- Threshold < 8 dB;
- AFC facility;
- Ability to process various video formats and frequencies.

The fundamental form of the receiver requires:

Input filter amplifier
This buffers the receiver first IF stage from the output of the LNB, providing both signal filtering and amplification. The filter removes image frequency signals from the LNB as well as matching between outdoor and indoor units. The amplifier itself not only gives signal gain but is designed to provide low intermodulation effects, a necessary characteristic for any system that is being used in a geographical location that has a number of higher power dbs transmissions arriving at the receiver.

Mixer systems
The mixer has to translate the signal from the 950–1750 MHz band down to an IF in the region of 500 MHz. In addition to this function it also has to prevent local oscillator signals being fed back to the antenna input; provide a switched local oscillator signal to allow channel selection and presetting from a local and remote position; and filter the IF signal to remove mixing frequency components before application to the demodulator system.

Demodulator system
The demodulator's basic function is to recover the transmitted baseband signals from the FM modulated carrier. The FM demodulator will require a constant input level and this is achieved by including the FM demodulator in an automatic gain control (agc) loop system. The loop itself will comprise an IF amplifier, whose gain can be controlled over a 40 dB variation in input signal level, followed by a filter, which determines the receiver selectivity and suppresses adjacent channel interference. The filter itself will almost certainly be a surface acoustic wave (SAW) filter with a centre frequency at the IF frequency and a bandwidth sufficient for the TV signal of 27 MHz.

The FM demodulator can be realised in a number of forms but is probably in the form a phase lock loop (PLL). This has the following advantages:

(a) It is realisable in digital form using integrated circuit techniques.
(b) It can provide an extension to the receiver threshold of up to 3 dB and so reduce antenna size or LNB performance.
(c) It is flexible in its performance capability, a vital necessity with the multiplicity of TV formats being used as well as new standards being introduced such as HDTV and MAC.
(d) Its response time, in terms of locking on to the input signal, is rapid, it has a high sensitivity and is simple to design and modify.

It can be seen from the above that the PLL fulfils the conditions set out at the start of this section.

9.5 SATELLITE MASTER ANTENNA TELEVISION (SMATV)

The alternative to dbs and individual satellite television reception is the use of communal receiving systems that either take in the satellite transmission and rebroadcast it, or distribute it by means of a cable network. It is intended as a reception point for low power satellite transmissions, which would require a larger antenna and more sophisticated receiving systems than the individual could afford or the local community would allow. The method of re-broadcasting the signal will probably be via the cable network, not only because of non-availability of terrestrial frequency allocations, but also because of interference by air transmission and the regulation of programme provision.

The advantages of SMATV are:

(a) The facility will remove the need for multiple installations of television receive-only systems (TVRO).
(b) The programme provision to the user can be controlled both in terms of access to the programming; the types of service provided; and the cost of service can be computed.
(c) It does not require any frequency spectrum.

The disadvantages of SMATV are:

(a) The cost of installing the system;
(b) Relative inflexibility in change of programmes.

The system basically comprises a head end and a distribution system, which would have a number of antennas looking at individual satellites.

9.6 TV UPLINKING

The programmes provided by any TVRO system are sent via the satellite from a central earth station. The uplinking station has to transmit sufficient power to the satellite so that the downlink to the consumer can provide the required quality of signal at the satellite receiver output. In performing this function it must take into account the environmental conditions, such as rain attenuation; the access characteristics of the satellite itself, where the power seen by the satellite has to be controlled to within strict limits; and prevent the possible interference with other satellites or terrestrial systems.

9.6.1 Uplink design

The definition of the uplink power is dependent on the information available to the designer. The parameters that are required are:

Receive power flux density at satellite

The RPFD received at the satellite assumes that the uplink station high power amplifiers are at saturation. This flux density is usually designated in dB/m^2. The flux density requirements are a function of the individual satellite and are given as a design parameter. The parameter is set assuming clear sky conditions and therefore any variation from this must be taken into account in any link design. The flux density is commonly known as incident power flux density (IPFD) and is usually different to the stated figure because of the use of a spreading factor which takes account of the fact that the actual satellite antenna, as seen by the transmitted signal, is not the same as the main beam antenna gain. Typical figures for the ECS1 and INTEL-SAT 5 satellites are $-84.5 \text{ dBW}/m^2$ and $-82.0 \text{ dBW}/m^2$.

Satellite antenna gain

It is necessary to know the gain per square metre of the satellite antenna receiving the signal from the uplinking station at the frequency of the uplink signal.

Uplink eirp

The designer must know the eirp from the uplinking earth station. The eirp takes account of the high power amplifier output, system losses and effective antenna gain. A typical uplink block diagram for a Ku Band system is shown in Fig. 9.8.

The modulated TV signal is upconverted from a 70 MHz IF to the transmit frequency in C or Ku Band. The basic equation must then take

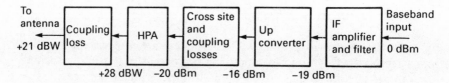

Fig. 9.8 Typical uplink for Ku Band earth station.

account of the losses due to waveguide runs, WG switches and couplers. This will then give the power available at the antenna feed which is simply the HPA output power at the output flange minus the losses in the coupling to antenna feed. It can be seen that it is essential that the losses are minimised. In addition system design must not just take account of black box parameters, such as HPA output and antenna gain, but the actual physical construction of the earth station, both in terms of the relative positioning of the antenna and equipment building and also in-building waveguide design.

The other major element in the eirp calculation is the antenna gain; this gain is basically a function of reflector size, frequency and antenna efficiency as derived from equation 9.6.

$$A_n = 10 \log \left(\frac{n \times \pi \times d \times f}{c} \right) \text{ dBi} \qquad (9.6)$$

where A_n = antenna efficiency
d = effective antenna diameter in metres
f = frequency in GHz
c = propagation velocity 3×10^8 m/s

The antenna will have an efficiency of between 50% and 75% and this would have to be related to the actual antenna chosen for the overall system design. It should be noted that the efficiency will be decided by the physical realisation of the antenna, surface accuracy, etc.

Free space path loss

The free space path loss at the uplink frequency has to be calculated to take account of the actual position of the earth station, relative to the satellite. The designer must take account of the true distance to the satellite and calculate the loss by the method shown in chapter 2 (Fig. 2.5). The attenuation in dB due to rainfall and any other atmospheric effects at the uplink frequency – this has been dealt with in chapter 2. The major factors to take into account are rainfall and gaseous attenuation and if possible the figure should relate directly to the area of operation. It is essential that the parameter value be set to the true value, rather than an arbitrary figure chosen to give additional margin. This will ensure that the HPA output power can be chosen at a minimal value and therefore lowest cost. In addition the value of the parameter will depend upon the availability required; for Europe the loss due to rainfall for a 99% availability level, will be 2 dB and for gaseous attenuation 0.2 dB. The rainfall figure increases considerably if 99.9% availability is required, i.e. 7 dB while the gaseous attenuation remains a constant.

Satellite G/T

The figure of merit for the satellite in dB/°K has already been referred to in chapter 4 for earth station receive systems and the same factors apply for the satellite itself. The actual figures of merit are available from the satellite operator's information.

Uplink carrier to noise C/No

This is the crucial relationship in determining the uplink system requirements. The C/No at the satellite can be determined with equation 9.7.

$$\frac{C}{N_0} = P_u - L_{fs} - L_g - L_r - dG_a + \left(\frac{G}{T}\right)_s \text{dBW/°K} \qquad (9.7)$$

where P_u = erip in dBW

L_{fs} = free space loss

dG_a = satellite antenna gain off beam

$\dfrac{G}{T_s}$ = satellite receiving G/T

L_g = gaseous attenuation

L_r = rainfall attenuation

For a typical system using an INTELSAT 5 satellite, operating in Ku Band at 14.5 GHz, providing an uplinking availability of 99% for the year uplink carrier to noise is as shown below:

$$\frac{C}{N_0} = 78.1 - 207.1 - 1.84 - 0.25 - 2.0 - 0.5$$

$$= -133.59 \text{ dB/°K}$$

The above figures assume a flux density at the satellite of -84.5 dBW/m^2 nominal and a dG_a factor of 2 dB. If the calculation is made using flux density parameters then the required equation is equation 9.8:

$$\frac{C}{N_0} = F_{ds} - G_a - \left(\frac{G}{T}\right)_s - k - L_r \text{ dB/Hz} \tag{9.8}$$

where F_{ds} = flux density assuming HPA saturation
$\quad\quad\quad G_a$ = satellite antenna gain
$\quad\quad\quad k$ = Boltzmann's constant ($- 228.6$ dBW/°K/Hz)

One other loss element that has to be taken into account is the possibility of fading, which can be up to 3 dB. This loss can be compensated for by the use of up-path power control which operates by monitoring the satellite beacon level and then increasing the eirp to counteract the fading level.

Chapter 10

Remote Sensing

10.1 INTRODUCTION

The monitoring of weather and the mapping and prospecting of the earth's surface has historically been a tedious business. Weather forecasting has been the bringing together of widely scattered information derived from balloons, weather ships and land instruments, followed by the interpretation of that data – not only the up-to-the-minute data but also the historical data that is available allowing the weather forecaster to take account of trends from the past. Mineral prospecting took place directly on the earth's service or at least by aerial survey. The advent of satellite communications led to the TIROS experiments, begun in 1960 by the USA, which monitored daylight cloud formations. This was followed in 1966 by ESSA–1 and ESSA–2, which were operational satellites but were limited in their application. It became obvious that what was needed was a system that could provide a complete global coverage necessitating both geostationary and polar orbiting satellites.

From these beginnings has developed a wider variety of uses for satellites over and above simple cloud monitoring. These uses included:

- Meteorological – for weather forecasting;
- Earth resources – for prospecting;
- Topographical data – for mapping;
- Ocean temperature and colour – for fishing resources etc.

Remote sensing systems utilise the behaviour and characteristics of the photons emitted from the sun to derive the data required. A photon is a discrete bundle of energy associated with the transmission of any electro-magnetic disturbance such as heat, light, radio and X rays. These photons go through two processes before arriving at the satellite:

10.1.1 Atmospheric effects on photons

The path of the photon through the atmosphere produces a number of changes in that energy, the effects being wavelength-dependent. They are listed below.

(a) The particles of dust, water, etc. have the effect of scattering the energy, the amount of scattering being dependent on the wavelength of the energy.

(b) The energy is absorbed by the atmosphere – again the attenuation is wavelength-dependent and absorption varies according to composition of the atmosphere. This means that at some emission wavelengths the energy of the photon is totally attenuated because of the characteristics of the gases in the atmosphere.

(c) The energy that will reach the earth's surface has had its characteristics altered not only by water absorption and the atmospheric gases but also by any elements emanating from the earth itself and in theory these could be identified by analysis of the photons.

10.1.2 Earth's surface effects on photons

The second phase of the photon modification occurs when the energy actually hits the earth's surface. This energy can then be:

(a) Absorbed by the various elements on the earth's surface, such as trees, soil, seas, etc. and the water content of those items affects the absorption rate.

(b) Reflected by the surface elements – reflection which is not simply dependent on the absorption rate of the elements but also their shape and geometry.

(c) The third factor that can be sensed is the energy emitted by the material at the earth's surface because of the incident energy arriving from the sun. In effect the earth's surface tries to maintain its energy equilibrium by emitting thermal infra-red energy. It should be noted that any object, whether at the earth's surface, or suspended in the atmosphere, will also exhibit this behaviour.

The above operational requirements have led to satellite, earth station and data processing developments, developments which many countries have undertaken, both for their own benefit and in hope of monetary return for the data derived from the remote sensing information.

10.2 METEOROLOGY

The current system is operated on a global basis with the various elements of the network being co-ordinated by the World Meteorological Organization. The major organisations/countries involved in meteorology have been:

- The National Oceanic and Atmospheric Administration (USA);
- The European Meteorological Satellite Organisation (EUMETSAT);
- The Japan Meteorological Agency;
- The National Remote Sensing Agency (India);
- The Indian Satellite Research Organisation (ISRO);
- The USSR.

The satellites used in meteorological applications are listed in Table 10.1.

Table 10.1 Meteorological satellites

Service date	Satellite system	Operator	Orbit type
1975	GEOS 75° E	NOAA	Geostationary
1975	GEOS 135° W	NOAA	Geostationary
1978	TIROS	NOAA	Near polar
1988	GMS 140° E	JMA	Geostationary
1987	METEOR 70° E	USSR	Near polar
1988	METEOSAT 0°	EUMETSAT	Geostationary

The meteorological satellites together provide nearly a full global coverage, with India providing the information from the Indian sub-continent. The satellites use multispectral radiometers which scan for the following emissions:

- 0.5–0.9 microns visible band;
- 5.7–7.1 microns infra-red absorption band;
- 10.5–12.5 microns thermal infra-red.

The nature of weather is such that the satellite must refresh the information regularly, because of the shift in weather patterns both rapid and long-term. In practice the METEOSAT satellites relay image information back to the European Space Operations Centre at Darmstadt, West Germany; this Centre then processes the information, archives it and distributes it to its customers. The satellites also receive data from a large number of land-based data collection platforms, which relay information up to the satellite to supplement the infra-red imaging. The basic information derived at Darmstadt is:

- Water precipitation index;
- Cloud cover data;
- Cloud top height data;
- Humidity values of the upper troposphere;
- Sea surface temperature.

The data is distributed to organisations that require the weather information. The global information gathered at the WMO centre is transmitted to the European Weather Centre at Reading in the UK, which produces global weather forecasts covering a period of three days. The frequency of operation for meteorology is shown in Table 10.2.

Table 10.2 **Meteorological satellite frequency bands**

Frequency (MHz)	Function	Data rate
402.0–402.2	Data collection	–
1670–2110	Satellite	166/333 kbits

10.3 REMOTE SENSING

10.3.1 Introduction

The remote sensing satellite has to carry equipment that can look at the earth's surface and detect radiation being emitted from it in two ranges of the spectrum, visible from 0.4–0.8 micron wavelength and infra-red ranging from 1.55–1.7 microns. The equipment used comprises:

(a) An optical system that can view the earth's surface at the correct sensitivity.
(b) A multiple element sensing array that can take the radiation received at each element and turn it into an electrical charge, which is then used to produce a voltage that can be converted into picture information. The sensor element is usually a photodiode that changes the light into an electrical charge which is stored in a MOS capacitor. It is important in sensor design that they have a high sensitivity and resolution, very low noise, and uniformity between sensors. In addition it is essential that the CCD devices have extremely high reliability as they have to operate over the full life of the satellite i.e. 7–10 years.

The satellite equipment is designed to scan the earth's surface in a series of blocks which cover a specific area. The basic system is shown in Fig. 10.1.

The resulting picture, built up from the satellite information is a series of areas which have the generic name of a pixel. It should also be noted that the information translated back to earth is given a colour that is not necessarily the actual colour of the object.

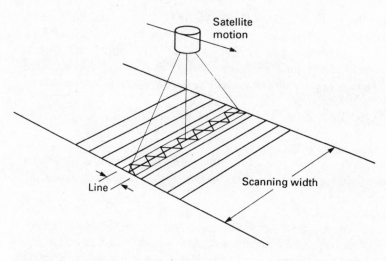

Fig. 10.1 Pushbroom scanning.

10.3.2 Remote sensing satellites

There are a number of satellites dedicated to remote sensing as shown in Table 10.3.

Table 10.3 Remote sensing satellites

Satellite system	Start of service	Country	Orbit
LANDSAT	1972	USA	Near polar
EOSAT	1984	USA	Near polar
SPOT	1986	France	Near polar
IRG	1988	India	Near polar
KOSMOS	Not known	USSR	Near polar

10.3.3 LANDSAT/EOSAT

The first major system to become operational was the LANDSAT service operated by the US Government until 1984. In 1984, due to political considerations, the Land Remote Sensing Commercialisation Act was passed in the USA, which effectively transferred the LANDSAT system to the private sector. The rules governing the use of remote sensing are laid down by the National Oceanic and Atmospheric Administration (NOAA).

The private company set up to continue the work of LANDSAT is known as the Earth Observation Satellite Company or EOSAT. It was financed by the US Government, an operation that allowed it to take over LANDSAT Satellites 4 and 5; develop new satellites LANDSAT 6 and 7 and ground control systems; and continue operation of the overall system. It is expected that the EOSAT organisation will become self-supporting by the sale of remote sensing information.

The LANDSAT satellites use a variety of sensors:

Multi-spectral scanner (MSS)

This measures the infra-red energy, both reflected in the band 0.7–1.1 microns and visible in the band 0.5–0.7 microns to a resolution of 80 m.

Thematic mapper sensor

This will cover the visible short infra-red range and the emissive thermal region of the spectrum to a resolution of 30 m for visible short range and 120 m for the emissive thermal band. This thermal mapper was used in the LANDSAT satellites, though for the L7 satellite this was enhanced to cover the 0.5–0.9 microns region, with a 15 m resolution for cartographic applications. The ETM will also view volcanic activity, soil erosion, crops, etc.

Wide field sensor (WFS)

This is designed to be compatible with the TM land data and ocean colour data, with a 500 m resolution, which is particularly useful for the fishing industry in determining the location of fish shoals by the ocean colour and the surface temperature of the sea.

The data collected by the satellite is relayed back to earth using X Band Carriers, which have a data rate of 15 Mbits for the MSS and 85 Mbits for the TM. The satellite is controlled from a ground control station at Norman in the USA. The data itself is downlinked to a number of earth stations scattered around the world. The area of the earth's surface that can be viewed from the satellite is limited, though this can be increased by pivoting the sensor to cover a wider area, in sections as wide as the non-pivoted swathe.

The information collected must of necessity be available to all and cannot therefore be transmitted in real-time to ground stations (assuming there are enough installed to allow the satellite to always 'see' one), as these stations may be restricted in their customer base. There are two methods by which the information can be independently relayed back to earth. In the first the data is recorded on board the satellite. For example in the later LANDSAT satellites, the recordings were carried out using wide-band data recorders that can store 7.5×10^{10} bits, thus providing worldwide coverage. The data is then downlinked to the Norman receiving station as the satellite passes within communication range. It is interesting to note that to save wear and tear on the on-board recorder this information is transmitted backwards to remove the need to rewind the recorder. Alternatively the data can be relayed from the satellite, which is in a near polar orbit, to a series of special satellites known as the Tracking and Data Relay Satellite System (TDRSS), which will give global coverage and relay the data to a main earth station.

10.3.4 Système Probatoire d'Observation de la Terre (SPOT)

The first operational satellite for the SPOT system was launched in 1986 to become Europe's first remote sensing satellite system. The major impetus to set up such a system came from France. SPOT is a public company owned by a number of French interests, with major holdings by CNES, MATRA, French banks and various organisations involved in mining, surveying and prospecting. In addition to these organisations there are a number of overseas branches, set up to exploit opportunities in the USA and Sweden.

The satellite is in a sun-synchronous orbit and can provide high resolution images. The data is received and stored in the satellite as it circles the earth and is then transmitted as digital data that represents both black and white and colour imagery.

The sensory system covers the spectrum in three bands:

- Green 0.5–0.59 microns;
- Red 0.61–0.68 microns;
- Near infra-red 0.69–0.89 microns.

The scanning system is of a very advanced design giving improved resolution, shorter acquisition times and higher signal-to-noise ratios; this performance is achieved by the use of a pushbroom scanning technique, where the charge coupled devices are mounted perpendicularly to the satellite movement and the scanning takes place by the natural orbital motion of the satellite. This gives a resolution of 10 m for colour and 20 m for multi-spectral coverage. The other advantage of this scanning system is that after a number of passes a particular scene is observed from slightly different viewpoints, giving three dimensional imaging of terrain. It provides much better information about mountain slopes, geological surveying, etc. at a much better discrimination than the LANDSAT system. The sensors themselves are steerable, allowing the satellite to see a 950 km wide strip of the earth. This ability gives the SPOT system a considerable advantage in that the satellite only needs approximately five days to produce repeat imaging. The speed of repetition is very useful from the users' point of view if they have a rapidly changing system such as fishing trawlers, forestry and farming.

The earth station system for the SPOT organisation provides world coverage with earth stations in:

- Canary Islands, France, Sweden;
- Bangladesh, China, India, Japan, Pakistan;
- Saudi Arabia;
- Australia, Brazil, South Africa;
- Canada.

The data received at the earth station is recorded and stored at the earth station itself. The recording system may use a high density data recorder which can store 30 000 bits per inch on each track; in practice the data is usually divided across multi-tracks in order to reduce the bit rate that each track has to cater for. The Thorn EMi high density recorders can handle data rates as high as 50 Mbits, with a packing density of 50 kbits per inch, a density achieved by using a special coding technique known as three-position modulation which represents three data bits by two flux transitions and the absence of a transition. The error rate of such equipment is 10^{-11} during recording.

The SPOT organisation is also a member of the EURIMAGE organisation which operates the EARTHNET system on behalf of ESA. The other members are Hunting Technical Services (UK), Telespazio SpA (Italy), DFVLR (W. Germany) and SATIMAGE (Sweden). This grouping's function is to sell remote sensing information derived from SPOT satellites, NOAA and ERS1 satellites.

10.3.5 Columbus space platform

The Columbus Development Program, which represents Europe's participation in the international space station contains three major elements:

APM or attached pressurised module
This will be a permanent part of the manned space station. This will be the environment for material science, fluid physics and life sciences experiments. The projected launch date is 1996.

MTFF or man tended free flyer
This is intended for long-term microgravity experiments that are essentially unattended except at the start and finish of the experiments. The projected launch date is 1998.

Polar platform
This is the earth observation element of the Columbus Platform. There are two options being considered:

(a) A platform, that would have components in common with the two other elements, using already proven technology to reduce the cost and design risks. The platform capacity is a total of 23 500 kg comprising a payload capacity of 23 000 kg and a support system of 500 kg, providing an average of 2600 W to the payload. It will have a life of six years, based on the amount of propellant.

(b) A platform that would be based on a SPOT spacecraft which has a payload of 1700 kg and provides an average power of 1700 W. The advantage of this approach is that the payload bus is a proven development.

The platform will be put into a sun synchronous polar orbit and will be of universal design to allow the remote sensing payloads to be fitted.

10.3.6 Indian remote sensing

The Indian space programme has, in addition to INSAT for communications, been involved in remote sensing since 1970, first using LANDSAT information. In 1973 they started building their first series of remote sensing satellites. The first, named Bhaskera 1 was launched in 1979, followed by Bhaskera 2 in 1981, both of which were used for land and ocean monitoring. The Bhaskera satellites were launched by a USSR rocket but with the second series of satellites, known as Rohini RS–D–2, the Indians used their own satellite launcher vehicle from a site at Sriharikota in southern India. These RS–D–2 satellites were very much an exploratory system in that they allowed India to gain valuable basic knowledge, not only of the Indian sub-continent but also of the needs and requirements of remote sensing satellites.

The control and direction of the Indian programme is the function of the Department of Space, which has within it the Remote Sensing Agency (NRSA); this agency has its headquarters in Hyderabad but has work carried out on remote sensing applications at the Space Applications Centre in Ahmedabad and at the Indian Institute of Remote Sensing at Dehradun.

The fourth phase of the Indian remote sensing programme became operational with the launch of the IRS 1A satellite, the first of the series that will take the project forward into the 1990s. The satellite specification is shown in Table 10.4.

The infrastructure of the Indian programme now comprises a central National Resources Management System with five Regional Remote Sensing Service Centres designed to facilitate the use of the received data, the first of which has been set up in Dehradun.

Table 10.4   IRS 1A satellite parameters

Parameters	Value
Orbit type	Circular (sun-synchronous)
Altitude	904 km
Stabilisation	3-Axes
Resolution	36.5 m/73 m
Coverage	148 km/150 km
Frequency bands	X/S band

10.4   COMMUNICATIONS FOR REMOTE SENSING

10.4.1   Introduction

In order to determine the frequencies to be used for remote sensing it is necessary to review the methods and instruments that are used for gathering the information required. In this section we will consider the forms of sensor and look in detail at some of the frequencies used and also some practical examples of satellite remote sensing systems.

10.4.2   Passive sensors

Passive sensors have been used for over forty years at first using aircraft but later, as satellite systems developed, specially designed satellites to carry out the remote sensing operations. The actual frequencies used depend upon the characteristics of the atmosphere which, as can be seen in chapter 2 (Fig. 2.10) has an attenuation characteristic that is dependent on the frequency, with particularly sharp frequency bands that have greatly increased attenuation. It is interesting to note in this context that the high attenuation areas at 60, 120 and 180 GHz are useful in remote sensing at the top of the atmosphere while in the attenuation areas in between the peaks the earth surface and atmospheric sensing can take place. The power received by the sensor is given by:

$$T_a = k \times B \times F_c \tag{10.1}$$

A simplified expansion of the above equation would give:

$$T_{a(s)} = T_0[1 - d_0(s) + T_s \, d_0(s)] \tag{10.2}$$

where T_a = antenna noise temperature °K

   B = receiver bandwidth in Hz

d_0 = optical depth in Nepers

T_s = atmospheric temperature at point S along the path

T_0 = surface brightness temperature (emission and scattering) °K

T_r = receiver noise temperature °K

The microwave receiver input signal is made up of a number of components and is given by:

$$T_{at} = T_a + T_0 + T_r \qquad (10.3)$$

The receiver has to distinguish between the thermal emission from the earth's surface, which looks like a noise signal; the noise that is gathered by the thermal emission as it travels towards the satellite; and the input noise temperature of the receiver. The optical depth is a function of path attenuation and the actual point along the path; it is defined by:

$$d_0 = \frac{A(s)}{4.35} \qquad (10.4)$$

where A = attenuation per km in dB

s = point on path in km

The equations indicate that to obtain the value of the parameter it is necessary to make more than one measurement. This means that simultaneous measurements have to be made at different frequencies in order to determine the actual value of the thermal emissions. The receiver has to have a low input noise figure, as low as 100 °K is possible at 12 GHz, and it can recover the information in the presence of unwanted signal path and input noise. The standard method of extracting the thermal emission and the input noise temperature is by integration of the total input signal. The error in this measurement can be reduced by decreasing the system input bandwidth or increasing the integration time, the latter being limited by the system requirements. The integration time can be given by:

$$t_i = \frac{k(T_0 + T_r)^2}{B \times \delta(T_0 + T_r)} \qquad (10.5)$$

The microwave sensors, as opposed to visible or infra-red frequency systems, have considerable advantage in that they are relatively immune to the effects of rain and clouds. The passive sensors have been used for measuring the atmosphere, the land's surface and ocean characteristics. The ocean qualities that are sensed are not simply at the surface but can be complex mixtures of surface water as well as under ice and water salinity, temperature, wind and snow. The measurement error, which is designed to be 10%, is affected by the frequency of measurement and by variations in the structure of the surface of the sea which produce an apparent change in dielectric constant. In Table 10.5 are listed various qualities of sea and optimum frequency of sensor operation.

Table 10.5   Frequency bands of sensor operation

Parameters	Frequency (GHz)
Salinity	1.5–3.0
Surface temperature	3–8
Wind	> 20

The above frequencies can be seen to overlap with fixed frequency services and care must be taken to prevent interference between systems. In practice frequencies up to 380 GHz are listed as preferred values, with specified bandwidths as shown in Table 10.6.

Table 10.6   Sensor frequencies for specific atmospheric measurements

Parameters	Frequency (GHz)	Bandwidth (MHz)
Salinity	1.4/2.7	100/60
Temperature	6	400
Water vapour	15	200
Surface ice	37	1000
Oil spills	90	6000
Temperature	118.7	2000
Water vapour	325.1	2000
Water vapour	380.2	2000

The above list is not exhaustive but shows some of the frequencies used. The emissions from the earth's surface can be measured by passive sensors but the signal received from the surface is proportional both to temperature and the effective emissivity and is affected by surface irregularities and scattering effects. Some of the recommended frequencies are shown in Table 10.7.

The atmospheric measurements differ in many ways from those shown in Table 10.7 in that use can be made of the edges of the opaque peaks in the atmosphere's attenuation characteristics, particularly for gaseous sensing. Table 10.8 shows the recommended frequency and bandwidth.

**Table 10.7 Sensor frequencies for earth remote
sensing**

Parameters	Frequencies (GHz)	Bandwidth (MHz)
Soil moisture	1.4/2.7	100/60
Snow	11	100
Snow	37	1000
Temperature	55	250
Snow	90	6000

Table 10.8 Gaseous sensing frequencies and bandwidths

Parameters	Frequency (GHz)	Bandwidth (MHz)
Rain	11	100
	15	200
	18	200
Water vapour	22.235	300
	24	400
	30	500
	37	1000
	183.1	2000
	325.1	2000
	380.2	2000
Temperature	55	250
	118.7	2000
Clouds	30	500
	90	6000
Nitrous oxide	100.49	2000
	125.61	2000
	150.74	2000
	175.86	2000
	200.98	2000
	226.09	2000
	251.21	2000
	276.33	2000
	301.44	2000
Ozone	110.80	2000
	184.75	2000
	235.71	2000
	237.15	2000
	364.32	2000
Carbon monoxide	115.27	2000
	230.54	2000
	345.8	2000
Chlorine oxide	164.38	2000
	167.20	2000

10.4.3 Active sensing

Active sensors have also been used for over forty years but differ from passive sensors in their ability to transmit an actual microwave signal to the object being sensed and receive back from it a reflected energy that allows the active sensor to determine the sensed object's characteristics. The disadvantage of active sensing is that the signal is subjected to double the atmospheric distortion as it travels down to the earth's surface and back to the satellite receiver.

The early use of active sensors relied on aircraft or balloons to carry the sensing equipment but since the 1970s a number of satellite payloads have been used to carry out active sensing experiments. The active sensor is used in three general ways; height measurements, earth surface contours and earth imaging.

Height measurements
This is perhaps the simplest of sensing areas in that it is essentially a radar system which transmits a narrow pulse and measures the reflected pulse. This technique has been refined over the years to improve accuracy by using such methods as pulse compression or time gating that give more information to the satellite receiver. It is obviously standard equipment for commercial and military aircraft all over the world, with the number being measured in tens of thousands and care has to be taken to ensure that terrestrial and satellite altimeter frequencies are not adjacent in the spectrum.

Earth surface contours
The contour of the earth is an important indicator for geological surveying as well as normal mapping applications. The way in which surface changes are measured is dependent upon the reason for the surface irregularity and the frequency being used. The type of instrument employed is known as a scatterometer which operates on the principle that a radar signal hitting the object will produce a back-scattered signal which can be analysed by the satellite. The type of measurements that can be made relate to:

- Soil water content;
- Vegetation;
- Wind velocity.

The water content of the soil can be obtained by measuring the dielectric constant of the soil, changes in which produce a change in soil reflectivity coefficient. The amount of vegetation will also affect the apparent soil water content level and it has been found that to reduce the effect of this, the incident angle of the sensing signal to the ground should be greater than 45°.

The measurement of vegetation is a complex process as the back-scatter can come from a variety of mechanisms within the target area. These are:

- Plant leaves;
- Stalk and leaf of plants;
- Soil irregularities;
- Scattering from the plants down to the earth.

In addition to this the signal power attenuation is inversely proportional to the cosine of the angle of incidence; proportional to the number of leaves etc. in a given volume. The shape of the scattering elements is also very important. The complexity of the measurement means that it is usual to take measurements at more than one frequency.

Wind velocity measurements are in general sea surface winds and are based on the fact that wave variation is a function of wind velocity and wind variations. The method of measurements will be described in more detail with the ERS 1 description in section 10.4.7 but essentially require multi-frequency, multi-beam measurements to give the actual value. The first satellite active sensing system was the S193 scatterometer aboard SkyLab from 1973 to 1974 which did not produce absolute values of wind velocity, the value being related to back-scatter, which is known as Bragg scattering.

Earth imaging

The use of sensors to produce high resolution images of the earth's surface began in the 1950s with airborne equipment and has developed for a wide variety of applications, such as geological surveying, current land usage and wave imaging. The major technique involved is the synthetic aperture radars or SARs. The most prominent of the recent satellite payloads is SEASAT, details of which will be considered in section 10.4.7.

Active sensor bandwidth

As will be seen in the systems descriptions the active sensing frequencies vary with application but whatever the type of sensor used, the system bandwidth is determined by certain of the instrument parameters and the resolution required of the image; the bandwidth is determined by equation 10.6.

$$B_s = \frac{c}{2R_r \cos \theta_a} \tag{10.6}$$

where
B_s = bandwidth in Hz

c = speed of light metres/second

R_r = resolution in metres

θ_a = angle of signal to its arrival point on earth

The above equation is also related to the effective radar pulse duration as in equation 10.7.

$$R_{pd} = \frac{1}{B_s} = \frac{2R_r \times \cos \theta_a}{c} \tag{10.7}$$

An idea of the resolution needed for particular types of sensor is given in Table 10.9.

Table 10.9 Active sensor parameters

Sensor	Resolution (m)	Swathe width (km)	Data rate (kbits)
HRT	10	48	12.10^5
IR radiometer	2000	2870	3300
Microwave radiometer	60 000	1350	5
Altimeter	.01	2	8
SAR	25	100	5.10^4

The high data rates needed meant that earlier satellites, such as SEASAT and LANDSAT use the 8 and 14 GHz bands for satellite to earth links, with LANDSAT utilising the TDRS 1 data relay satellite as a means of returning information to the ground station.

10.4.4 Frequency allocations for remote sensing

Introduction

The considerations that have to be made on frequency allocations are ones concerning the type of measurement to be made and the effect of the remote sensing transmissions on the existing satellite and terrestrial services. The ITU has carried out considerable work in defining the boundaries of operation and the allowable levels of interference. In practice each satellite differs in its frequency and modulation schemes depending upon location of ground stations etc. but the ITU allocations provide for two bands for telemetry data from remote sensing satellites; 2.2–2.29 GHz for downlinking to the earth.

Passive sensing

The ITU has considered a range of frequency bands and has come to the following broad conclusions:

1.4 GHz Band
This is not universally usable for passive sensor operation, due to the high signal levels involved, when the satellite can 'see' more than five terrestrial

terminals, fixed or mobile. This means that over most of the CONUS, European and Japanese regions this frequency cannot be used and obviously as the fixed and mobile services spread the barrier to the use of 1.4 GHz will also spread.

10 GHz Band
This is limited to the fixed satellite service and by the eirp being transmitted, which must not exceed 38 dBW.

15 GHz Band
This is not feasible at all in terms of remote sensing but can be used for radio altimeter work.

21 GHz Band
This contains no problems for sharing with fixed services but if mobile services are involved then there is a limitation on the number of mobiles that can be 'seen' by the satellite at any one time.

37 GHz Band
This offers some sharing capability with fixed and mobile services but unacceptable loss of data will occur, due to interference, if the number of terrestrial terminals is greater than 20.

50–70 GHz Band
There are no problems with this band of frequencies when considering geostationary platform operation but there is a likelihood of some loss of service when low-orbit satellites are used.

Active sensing

For active sensing the level of interference will depend upon the output levels of the SAR etc. and these can be kept low in order to prevent problems arising.

10.4.5 Meteorology

Introduction

The definitions of the frequencies used for meteorology are separate from earth resources measurement though there are overlapping functions between the two systems. The main areas of frequencies used are in the band 8, 9, 10. The characteristics of the low-orbit satellite are determined by the fact that the meteorological information, associated with the satellite, is derived from three major sources:

(a) Sensors housed in the satellite that operate at visible, infra-red and radio frequencies;
(b) Collection points on the earth's surface which transmit data up to the satellite. This is then transmitted down a main earth station used for processing it;
(c) Search and rescue data derived from ships and aircrafts.

The satellite transmits down to a variety of earth stations depending upon the information to be sent:

(a) A central processing earth station, which also acts as a command and control centre;
(b) Automatic picture transmission (APT) receive earth stations, where the information is sent in real-time;
(c) High resolution picture transmission (HRPT) receive earth stations also operating in real-time;
(d) Search and rescue data which is sent to the rescue co-ordinating ground station.

The frequencies of two low-orbit systems, NOAA (USA) and METEOR (USSR) are shown in Table 10.10.

Table 10.10   Low orbit meteorology satellite frequencies

Information	Frequency (MHz)	
	NOAA	METEOR
Sensor data	1698/1702.5/1707	466.5
S and R (down)	1544	1544
S and R (up)	121.5/243/406.5	121.5/243/406.5
APT	137.5/137.62	137.15/137.3/137.4/137.5
HRPT	1698/1707	
Data collection	401.65	

Geostationary systems operate in a different manner and here information is derived and controlled by:

(a) Active and passive sensors which measure a variety of weather parameters such as cloud characteristics, sea and earth temperatures, atmospheric and storm conditions;
(b) Weather facsimile and ranging signals transmitted from the command and acquisition centres;
(c) Data collection from ships, aircraft and buoys.

Data is received from the satellite by two means:

- A link from the satellite to the main control earth station.
- Direct transmissions down to individual user small earth stations.

The basic frequencies used are shown in Table 10.11.

Table 10.11   Geostationary meteorology satellite frequencies

Function	Europe METEOSAT	Frequency Japan GMS2	USA GOES	India INSAT
Telemetry frequency (MHz)	1675.929	1694	1694	–
Number of data collection channels	66	33	66	33
DCP frequency S/E	1675.281	1694.5	1694.5	4000
Raw image frequency (MHz)	1686.333	1681.6	1681.6	4000
Raw image data rate (kbits)	166	> 2000	28 000	400
Dissemination Ch F (MHz)	1691	1691	1691	–
Ranging frequency (MHz)	1691/1694.5	1684/1688.2 1690.2	GMS2 + 2209.086	–

Interference with meteorological transmissions

The amount of interference from other sources determines the use that can be made of the 400 MHz band. In this band interference can come from other meteorological aids known as radio sondes as well as radio theodolites; this interference from radio sondes can be at both 400 MHz and 1670 MHz.

The radio sondes are launched at regular intervals four times per day and transmit information on weather conditions at output powers between 0.5 and 1.0 W, using a monopole antenna which has at least 2 dBi gain. This performance will present no interference problems under normal conditions but for the radio theodolite case, which operates at a higher eirp of 20.8 dBW, which will give an RPFD of − 159.2 dBW compared to a DCP level at the satellite receiver of − 162.4 dBW it presents interference problems. The various possibilities are considered in CCIR Recommendation 362–2.

10.4.6   Meteorological earth terminals

Small METEOSAT earth stations
METEOSAT satellites can transmit directly to data users; the earth station parameters are shown in Table 10.12 as well as the future parameters that will allow the data to be received by a Standard A INMARSAT terminal.

The signal at the earth station receiver can be calculated from equations 10.8 and 10.9.

Table 10.12   METEOSAT earth station parameters

Parameter	Value	
	Current	Required
G/T dB/°K	+ 2.5	− 4
Antenna diameter (m)	2.0	0.9
Noise temperature °K	400	400
PFD dBW/m²	− 141.8	− 141.7
PFD/4kHz	− 145.3	− 145.3
S/N (30 kHz) (dB)	> 12	> 12

$$P_r = A_a \times \lambda^2 \times \frac{P_{fd}}{4\pi} \qquad (10.8)$$

and

$$\frac{S}{N} = \frac{\dfrac{A_a}{T} \times P_{fd} \times \lambda^2}{4\pi k B} \qquad (10.9)$$

where A_a = antenna gain

λ = input frequency

P_{fd} = power flux density

k = Boltzmann's Constant

B = bandwidth in Hz

METEOSAT TT and C Station for data acquisition
The basic parameters are shown in Table 10.13.

Table 10.13   METEOSAT TT and C station parameters

Parameter	Value
Antenna gain (dBi)	45
Antenna beamwidth (degrees)	0.8 (3 dB point)
System noise temperature (°K)	115
Frequency (raw data) (MHz)	1686.833
Rx bandwidth kHz	5400
Pfd dBW/m²	− 142.8

10.4.7   Data collection platforms/TDRS

Introduction

One important element in the overall system is the collection of data, this has already been referred to earlier and is achieved by:

- Land-based data collection units (DCU);
- Satellite receiving the information from the DCUs;
- Satellite transmitting down to a TT and C;
- Satellite transmitting direct to the user.

The above applies to both geostationary and low-orbit satellites and the basic characteristics of two such systems are shown in Table 10.14.

Table 10.14   Data collection platform characteristics

Parameters	System Argos	DCP
Orbit	Low	Geostationary
Altitude	830 km	–
Emission	Intermittent	Intermittent
Repetition rate	100–200 s	–
Carrier frequency	401.65 MHz	402 MHz
Pt W	3	10
Bit rate bit/s	400	100
Rx signal level	– 120 dBW/m²	– 145 dBW/m²

Tracking and data relay satellites (TDRS)

The purposes of the TDRS is to improve coverage of low-orbit satellites and thus reduce the number of earth stations required to give global real-time coverage. The TDRS is a NASA designed system. The first satellite was launched in 1983, but due to launch problems was only able to be put into a low-earth orbit. The second satellite was aboard the Challenger shuttle that exploded in 1986; a third satellite was launched in 1989 to replace the lost satellite and the original 1983 satellite has been repaired. Thus the TDRS system can now provide a service to LANDSAT and the space shuttle as well as C Band satellites and 19 other low-orbit satellites. The system does not provide full global coverage, but only one part of the Indian Ocean region remains uncovered by the satellite transmissions. An idea of the cost advantages of the TDRS can be gauged by the fact that NASA has been able to reduce its satellite earth station network from 23 down to four systems.

The TDRS satellite operates in a variety of frequency bands, depending upon the function; for low to medium data rates, up to 250 kbits. The data is modulated onto a 2287.5 MHz carrier, which is transmitted from the low-orbit satellite to the TDRS satellite which, by the use of spread spectrum techniques, is able to deal with up to 30 satellites. The combined signal is transmitted to the TDRS earth station on a 13.5 GHz carrier, where each of the 30 spacecraft signals are separated and processed.

The frequencies used in the satellite are shown in Table 10.15.

Table 10.15 TDRS frequencies

Frequency (MHz)		Value	Data rate
TDRS–LOS LDR		2106.4	< 250 kbits
TDRS–E/S LDR	Band 9	13.5 GHz	< 250 kbits
E/S–TDRS LDR		14.8 GHz	< 250 kbits
LOS–E/S LDR		2287.5	< 250 kbits
TDRS–LOS HDR		2025–2110	< 1 Mbit/s
LOS–TDRS HDR	Band 9	2200–2290	< 1 Mbit/s
E/S–TDRS HDR		14.68/14.72 GHz	< 1 Mbit/s
TDRS–E/S HDR		13.768/13.698 GHz	< 1 Mbit/s
TDRS–LOS HDR		13.775 GHz	< 1 Mbit/s
TDRS–E/S HDR		13.53/13.93 GHz	< 1 Mbit/s
LOS–TDRS HDR	Band 10	15.0 GHz	< 1 Mbit/s
E/S–TDRS HDR		14.625/15.2 GHz	< 1 Mbit/s

10.4.8 Operational earth exploration satellites

ERS–1

The European Space Agency (ESA) intends to set up a fully operational system that will provide a commercial service into the 1990s. The pre-operational satellite, ERS–1, was launched in 1989, though the satellite payload and characteristics had already been fully characterised. The ERS–1 satellite is intended to prove not only the satellite payload but also the ground stations necessary to control the satellite and receive the sensed information for processing.

The ERS–1 satellite payload is:

- An active microwave instrument (AMI), which contains a radar altimeter, wind scatterometer and a synthetic aperture radar;
- An along the track scanning radiometer (ATSR);
- Microwave sounder;
- Precise range and range rate experiment (PRARE);
- A laser retro-reflector.

The altimeter operates in Ku Band at 13.8 GHz, with a bandwidth of 390 MHz. The altimeter accuracy depends upon the earth's surface elements, over water it will be ± 10 cm, over ice it will be ± 40 cm. The altimeter accuracy allows ocean wave variations to be measured and the altimeter can be switched from ice to ocean measuring mode. This change can take place in less than five seconds without a loss of data.

The wind scatterometer mode of the AMI package interleaves with the wave measurements and operates on sideways looking antennas which give three beams, mid, fore and aft. The signal inclination is 45°, giving an accuracy of ± 10% for wind speeds in the range 4 m/s to 24 m/s, as well as the wind direction, to an accuracy of ± 20%. The scatterometer measures the radar cross-section of the sea for each 25 km square block of the swathe.

The synthetic aperture radar uses a rectangular antenna, along the line of movement, to send a signal down to the earth's surface at an angle of incidence of 23°. The SAR operates in two modes, imaging and wave, though both modes use identical radar geometry, which is shown in Fig. 10.2.

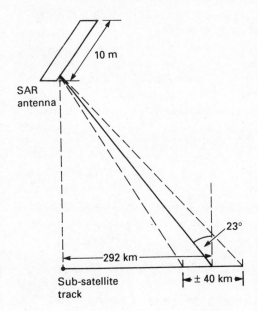

Fig. 10.2 SAR geometry.

The geometry of Fig. 10.2 is determined by the angle of slant of the radar transmission. This slant gives an angle of incidence of 23°, at 292 km from the sub-satellite point on the earth's surface. The antenna beamwidth of 5.4° cuts a swathe of ± 40 km. The radar footprint is 4.5 km at 5.3 GHz and this would be the spatial resolution in azimuth if it were not for the application of aperture synthesis, which gives a spatial resolution of 30 m, a result achieved by producing an apparent increase in effective antenna length due to the signal processing. This measurement is made for a ten-minute period during each orbit.

The SAR wave mode senses an area of ocean 5 km square, with a resolution of 30 m. The SAR data is transmitted in real-time on an

8140 MHz frequency carrier using PSK modulation with a data stream up to 105 Mbit. The measurements are made at intervals along the orbital arc, intervals which are selectable at either 200 km or 300 km. The satellite processing produces a two-dimensional model of the wave patterns.

The ATSR utilises a microwave radiometer operating at 23.8 and 36.5 GHz and a passive infra-red radiometer which measures sea surface temperature as well as water vapour and cloud formations in the atmosphere. The two sub-systems can be operated simultaneously or independently over long periods of time in order to achieve thermal equilibrium.

The PRARE experiment is concerned with orbital accuracy and is intended to measure the satellite radius to within ± 10 cm. The orbital information is transmitted down to the Darmstadt Control Centre.

The ERS–1 experiment is also intended to prove out the ground station system; this comprises four major earth stations:

- Kiruna, Sweden, which provides acquisition capability and SAR fast delivery processing and dissemination. It is also the TT and C station for the whole operation, acting as the Mission Management Centre interface.
- Fucino, Italy, which is as Kiruna apart from its TT and CV capability.
- Gatineau, France, provides data acquisition and fast delivery processing.
- Maspalomas is similar to Gatineau.

The above will be supplemented by national and mobile systems, while the Mission Control is situated at Darmstadt in West Germany. This station takes in the information from Frascati in Italy, the Central User Service, which logs all user requests and essentially acts as a project management facility for ERS–1.

Marine observation satellite (MOS–1)

Japan launched its first earth observation satellite on 19 February 1989. The satellite, MOS–1, weighs 730 kg and has an orbital period of 103 minutes, with a perigee height of 903 km and an apogee height of 917 km. The satellite payload is intended to prove out remote sensing technology for the following systems:

(a) A multi-spectral electronic self scanning radiometer (MESSR) which uses visible and infra-red radiometers, with pushbroom (CCD) scanning. The MESSR senses sea surface colour over the range of 0.51–1.1 microns in four bands. The data is derived from a 100 km swathe and is sent back to earth on an 8150 MHz carrier MSK modulated by an 8.78 Mbit data stream, with a 5 W transmitted carrier power.

(b) A mechanical scanning radiometer operating at visible and infra-red to measure the sea surface temperature. The VTIR senses a swathe of

1500 km and sends information back to earth on an 8350 MHz/5 W carrier.

(c) The satellite payload also has a data collection system (DCS), with terrestrial data collection platforms (DCP) transmitting to the satellite at 401 MHz and then transposed to the 1702.4848 MHz frequency for transmission of the data back to the Earth's Observation Centre at Tanegashima in Japan.

SPOT 1 Satellite

The SPOT remote sensing satellite became operational in 1986 in a near-polar orbit at an altitude of 822 km, with an orbit time of 101.3 minutes. The satellite is designed so that, even though it completes a scan of the earth every 26 days, it actually can complete observation of the same equatorial area eight times in that period of 26 days thus allowing the operator to focus more closely on particular areas. The satellite monitors forestry and soil utilisation, explores mineral deposits and carries out new 1:100,000 mapping and revises existing mapping. The measurements are carried out using:

(a) A high resolution panchromatic sensor;
(b) A three-channel multi-band sensor.

The satellite has two picture packages made up of (a) and (b) independent of one another. The data from the package is recorded on 50 Mbit video recorders, though the recording is carried out sequentially so that as one recorder records the other is replaying its previous data. The data from the VTR or picture taking packages is relayed to the ground station on a 8025–8400 MHz, 20 W, carrier. The direct output will be PCM encoded at a bit rate of 25 Mbits.

SEASAT Satellite

The SEASAT satellite was intended for a specific mission, to measure ocean characteristics. The satellite was launched in 1978 into an orbit that took 103 minutes to complete, covering 95% of the earth's ocean surface every 36 hours. The satellite mission only lasted three months, due to a sub-system power failure, though not before most of the required data had been gathered. The measuring instruments on the satellite were:

- Radio altimeter;
- A wind field scatterometer;
- A synthetic aperture radar.

The payload was meant to measure ocean characteristics such as wave height, current patterns and ice fields as well as wind speed and direction.

Chapter 11

Satellite and Optical Communications

11.1 INTRODUCTION

A chapter on optical communications might seem to be somewhat out of place, however, the major competition that satellite communication will face in the future will be optical fibre systems, especially for long distance point to point communications. The increase in optical fibre links is a very important pressure on satellite communications, not only in terms of communications costs, but also in the direction that satellite communications will take in the future. In order to remain a viable alternative to optical fibre the satellite operators have not only to reduce their costs but also to increase their range of services and communications options.

The use of cable for communications is far older than satellite and in the early days of satellite communications, it was considered that satellites would, with their ability to provide wide area coverage, signal the demise of cable communications, especially transatlantic underwater cable. The major advantages of satellites were seen as:

(a) Point to multipoint capability;
(b) Flexibility, both in the user's ability to change the format of the system and in the type of service offered;
(c) The system is, to all intents and purposes, insensitive to distance, assuming a geostationary orbit;
(d) Ease of installation, with a system being easy to set up in a relatively short space of time, without the problems of digging up city pavements or laying underwater cables.

The reality of the situation is that after more than 25 years of satellite communications the cable communications sector, both coaxial and optical fibre, is thriving. In the case of transatlantic cables for instance the traffic is divided evenly between cable and satellite. The increase in optical fibre communications has occurred for the following reasons:

(a) The rapid development of the optical cable itself, offering as it does wide bandwidth, reasonable cost, ease of handling, small receive/transmit terminals and freedom from electrical interference.

(b) The move towards terrestrial digital communications which increased the complexity of satellite communications and necessitated more costly systems such as TDMA.

(c) Satellite technology, for all its advanced capability, in fact took a time to react to the pressure of users' requirements in terms of services.

(d) The regulation of satellite communications, though necessary to ensure that all countries could utilise such systems as INTELSAT and INMARSAT, brought with it a rigidity of development. It should be said that this is not true in areas of satellite use which give the user a positive advantage over other forms of information carrying, such as television distribution. An extension to this can be seen when high definition television becomes a reality, for this can only be 'broadcast' via satellite due to the bandwidth required for its transmission. This is not to say that optical fibre cannot provide that type of interconnection but not for the same cost or with such ease of implementation. It has been estimated that to cable the UK with optical fibre would cost £20 billion at 1989 prices, a task beyond any commercial company.

One further element that has to be considered, when comparing the two forms of carrier system, is the reliability, both inherent in the equipment and as an operational factor. The lifetime of satellites is constantly being extended, not only by improved component and sub-system performance, but by extension of the satellite station-keeping ability. For optical fibre cable the long-term reliability is unknown, in operational terms, from the point of view of the cable itself, but it is certainly known that optical fibre cables can be dredged up, cut through or even attacked by sharks, causing considerable loss of traffic and revenue, as well as the unknown element of the effect of water upon the cable itself. It is very unlikely that satellites will fail in service, though failure occurs more regularly during the delicate operation of launching the satellite, putting it into the correct final orbit and checking out its sub-systems.

A characteristic, often of importance to the long-term planning, is the question of security of communications both physically and electronically. In this respect both forms of carrier system have advantages and disadvantages:

(a) Both satellite and cable can be attacked physically, though it is obviously more difficult to damage a satellite than cable.

(b) Electronically the cable system is very secure and is relatively immune to electronic jamming. The satellite however, can be jammed or damaged by the use of high power signals transmitted towards it. This form of jamming can possibly be overcome by a change of transponder frequency and an increase in the communication power transmitted to the satellite.

(c) The satellite transmissions can be intercepted and the information gathered for the purpose of military or commercial espionage. This can be overcome by use of sophisticated encryption systems.

11.2 OPTICAL FIBRE COMMUNICATIONS

11.2.1 Introduction

The idea of using light to send messages was put forward as far back as 1880 by Bell but it was not until the invention of the laser, in the early 1960s, that optical fibre cable communications became possible. This was because lasers gave the designer a coherent light source that could be modulated to allow transmission of information. In 1966 scientists in STL put forward the idea that light could be guided along glass fibres, with considerable advantages over ordinary cables. The first production fibres became commercially available in 1970. These had only one-tenth of the transmission loss of coaxial cables, while modern optical cables have a loss of approximately 1 dB per km on a 140 Mbits system, compared to 30 dB for coaxial cable.

Optical fibre works on the principle that if light is transmitted along a fibre then the use of a central core, which has a higher refractive index than the outer cladding, will ensure total internal reflection. There are two types of fibre available, multi-mode and monomode; the multi-mode fibre allows many light waves to travel along it, giving higher coupling efficiencies between fibre and sources/detectors. It is however, more expensive than monomode and has distortion effects due to multi-mode transmissions. In monomode only a single ray of light is allowed to travel along the core, with low losses and distortion and it is, therefore, used in many long haul systems.

The signals travelling along an optical cable are subject to attenuation because of three main factors:

- Absorption of light due to impurities in the glass;
- Scattering due to the glass structure and radiation;
- Interfacing with the opto-electronics and interconnection losses.

The main advantages of optical fibre compared to coaxial cable are:

- Very low attenuation;
- Wide bandwidth of up to 1000 MHz;
- Immunity from electrical interference;
- No cross talk between cables, giving improved quality of performance;
- Low weight and small size, allowing ease of handling.

11.2.2   Cable design

The optical fibre cable is in itself fragile and requires protection if it is to be installed either below ground or under the sea. Its environment can subject the cable to water, frost, soil corrosion and physical damage and therefore the outer protection to the optical fibre must remain unaffected by any of these external factors. At the same time the finished cable must be easy to handle for installation and repair, implying that the overall cable is small in physical size and reasonably flexible. The cable design itself should not stress the individual fibres and therefore the preferred system for cables, that are installed underground or under water, will be the non-rigid system that is pressurised to prevent the ingress of moisture but does allow the fibre freedom to flex inside the outer protection. The result of research on practical cable has shown that the best results can be achieved if the cable is not only flexible but is protected from external radiation sources that can seriously impair the cable performance.

11.2.3   Optical sources and detectors

The optical transmitter is either a light emitting diode (LED), when a multimode fibre or low bit rate system is required or a laser source when monomode fibre is used. Obviously the development has had to keep pace with that of the optical fibre, which means that the noise characteristics must be such that there is no distortion added to the signal, impairing transmission. The modulation capability must be such that the source can deal with the very high data rates now in use. In addition it is important that the system be as efficient as possible. Laser sources can provide low noise outputs and have a modulation capability of over 565 Mbits. This performance is particularly good when a window type laser is used, which has a bandwidth of more than 10 GHz. It is also essential that in this type of transmission optical reflections do not occur anywhere along the link as these produce periodic noise spikes. Reflections occur at discontinuities along the fibre such as repeaters or cable joins and where fibres have to be joined together the reflections are minimised by the technique of fusion splicing of the fibres.

The photodetector detects the transmitted signal by converting the variations of optical intensity into an electrical signal. In addition they must be able to handle the information rate used on the link; have low shot noise; and be physically compatible with the fibre size. The type of detector most used for long distance transmissions is the avalanche photodiode, while for short links the pin diode is more cost-effective.

11.2.4 Wavelength multiplexers

The use of broadband optical networks will involve 140 Mbits and 565 Mbits line systems and although this may be sufficient for a particular network it will often be necessary to bring together a number of separate digital hierarchies at say 34 or 140 Mbits. One method of achieving this is to use wavelength division multiplexing (WDM). The basic system can be illustrated by reference to Fig. 11.1, which can apply to any line system and bit rate.

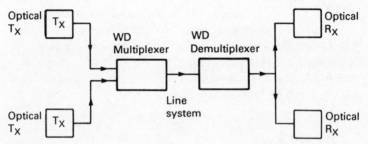

Fig. 11.1 Wave division multiplexing system.

The WDM has to provide both multiplexing and demultiplexing functions and often these are realised by two different techniques; the multiplexer is in essence a coupler which couples in a number of different signals to produce a single output. The techniques available for this process are:

(a) Dielectric thin film filters, suitable for wavelengths up to 500 microns.
(b) Gratings used as optical filters as shown in Fig. 11.2.

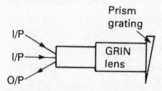

Fig. 11.2 Wave division multiplexing using GRIN lenses.

This WDM can operate to a wavelength of 1 nm. The term GRIN stands for GRaded INdex miniature lens.

(c) Directional couplers which can operate over the same wavelength as the dielectric film filters.
(d) Fused couplers, produced when two monomode fibres are fused together allowing the two signals to couple together with a very low attenuation (> 5 dB). The performance is controlled by the length over which the fibres are fused together, a length that can be used to give a very high coupling factor at one wavelength while rejecting all other signals.

(e) Waveguide or periodic filters, which use an interferometer known as Mach-Zehnder.

11.2.5 Wavelength demultiplexing

The demultiplexing of optical signals can be accomplished by the use of optical filtering. The optical system is shown in Fig. 11.3.

Multiplexed I/P $f_1 + f_2$

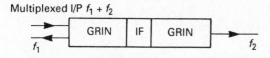

Fig. 11.3 Wave division demultiplexer.

The optical system consists of two GRIN lenses that are separated by an optical filter; the filter allows f_2 to pass through and concentrates the signal towards the outgoing fibre, while at the same time reflecting the f_1 signal, which is concentrated by the lens towards the f_1 output.

11.3 OPTICAL FIBRE VERSUS SATELLITES

It is undoubtedly true that the future of point to point communications, particularly long distance intercontinental systems will be dominated by optical fibre. For terrestrial systems the picture is less clear and is more determined by the user's actual requirements and the distance to be covered than any other consideration. It is also true that for point to multipoint communications the satellite solution will be the most effective.

The major area of cable use is in transoceanic communications, with the first coaxial cable, named TAT1, being installed in 1956. It connected the USA to continental Europe. This use of cable has grown steadily and even coaxial cable underwent considerable development due to pressure from satellites. The continual growth in demand for capacity ensured that cable and satellite services ran side by side to provide the necessary services.

The major drawbacks with cable systems are:

(a) Repeaters are required along the cable to overcome the attenuation of the cable, for example TAT7, which offered a capacity of 4000 circuits, requires a total of 660 repeaters.
(b) Although the number of repeaters can be reduced by reducing the cable attenuation, this implies the use of larger diameter cables, with the consequent effects on cable manufacturing costs, difficulty of laying larger cables and long-term maintenance.

(c) If the cable size is fixed then longer distances can only be accommodated by increasing the number of repeaters, again with a consequent increase in overall cable costs, more difficulty in maintenance due to the shorter cable links between repeaters and a reduction in link reliability.

(d) Cables can be subject to predatory attack, damage through dredging and fishing and are difficult to install. The early transatlantic cables, up to TAT8, provided analogue circuits; beyond that the move to digital communications continues.

Current transatlantic cable systems

Table 11.1 shows the cable systems now in service.

Table 11.1 Transatlantic cable systems

Cable	Route	Capacity (voice circuits)
TAT1	Newfoundland/Cornwall	48
TAT2	Newfoundland/France	48
TAT3	Tuckerton/Cornwall	138
TAT4	Manahawkin/France	138
TAT5	Greenhill/Spain	845
TAT6	Greenhill/France	4000
TAT7	Tuckerton/Cornwall	4200
TAT8	Tuckerton/Cornwall	296 Mbits
CANTAT1	Newfoundland/Scotland	80
CANTAT2	Beaver/Harbour/Cornwall	1840
COLOMBUS1	S America/N Africa	1840

These are the main transatlantic cables. It should be noted that for cables that terminate in the UK, the continental connection is established by spurs across to Brittany. The further projection of transatlantic cable to 1991 should see TAT9 in place which will provide 560 Mbits of digital capacity.

Communication growth

The factors affecting the growth of optical fibre cable are both commercial and technical in that by 1988 there were 11 337 voice circuits and 4625/64 kbit channels available across the Atlantic, plus satellite capacity of 53 500 voice circuits derived from INTELSAT 3 to INTELSAT 6 satellites. This capacity, assuming that voice multiplication techniques are also used, will cater for the projected growth to 1997, as shown in Table 11.2.

Trans-Pacific cables

In the Pacific there is a similar situation, i.e. a rapid growth in the use of cable, with acceleration due to optical fibre as shown in Table 11.3.

Table 11.2 Projected transatlantic capacity requirements

Year	Voice circuits (1000)
1985	42
1988	65
1991	90
1994	150
1997	200
2000	300

Table 11.3 Pacific transoceanic cable systems

Route	Capacity (circuits)
Hawaii/mainland US	845
Hawaii/Canada	3068
Hawaii/Australia	3068
Hawaii/Guam	1445
Singapore/Hong Kong	1380
Singapore/Australia	1620
Singapore/Philippines	1620
Taiwan/Philippines	600
Taiwan/Okinawa	480
Taiwan/Guam	845
Taiwan/Hong Kong	480
Japan/Okinawa	1960
Taiwan/Philippines	140 Mbits
Guam/Philippines	140 Mbits
Guam/Hawaii	280 Mbits
Mainland US/Hawaii	280 Mbits

The above is a static picture and takes no account of the future plans, which will add 840 Mbit cables between Japan, Hawaii, Australia and Guam in a ring installation.

The first generation utilises the following:

- Monomode fibre (1.3 micron wavelength);
- Repeater spacing > 40 km, which assumes a cable loss of approximately 0.5 dB;
- In-service protection in repeaters;
- Equipment with a lifespan of at least 25 years;
- MTBF > 10 years.

The second generation systems will utilise new developments to give the following characteristics:

- Monomode fibre (1.55 micron wavelength);
- Repeater spacing > 80 km, which assumes a cable loss of 0.25 dB;
- Operating frequency 140 Mbits.

11.4 FACTORS INVOLVED IN COMPARING SATELLITES AND OPTICAL FIBRE

11.4.1 General

The factors that must be considered when comparing satellite and optical fibre transmissions are not just technical, account must be taken of the total costs involved in using either system and any other factors that impinge on the use of the system. These are:

(a) Capital cost of terminal equipment;
(b) Capacity of the system;
(c) Services required over the communication link;
(d) The capital costs associated with the setting up of the service. In the case of the satellite system this would be the cost of building the satellite, launching it into orbit, maintaining it in its fully operational mode while in orbit plus earth station costs. For the optical fibre cable, capital costs involve producing the cable and the repeaters, buying wayleave rights, laying the cable, either as a terrestrial or as a transoceanic link and installing the terminal equipment.

11.4.2 System operating costs

The items listed in section 11.4.1 relate to the system start-up costs and account must be taken of the operating costs which are:

(a) The cost of financing the original investment, which will involve the user in interest charges on the capital employed and capital repayment.
(b) The cost of the system once it has been installed, involving consideration of the daily expenditure to ensure that the system availability is maintained at the required standard. For satellites this involves leasing the satellite capacity, the control centre running costs, earth station equipment maintenance, power costs for running the earth station and overall building maintenance. For optical fibre this is much more variable in the sense that the cable-repeater maintenance is probably only required on a 'need' basis if catastrophic interruption of service occurs. Equipment failure is much more serious, even with the use of equipment redundancy, due to the high cost of cable-repeater recovery.

11.4.3 Future proofing

There is a further element that must be taken into account in comparing satellite and cable costings and that is the possibility of further enhancements. Obviously the satellite cannot be changed while in an operational mode, though the space shuttle can now recover satellites for refurbishment and upgrading. In general however, the satellite must be used to its maximum capacity over its expected lifetime of ten years.

The capacity can be increased by earth station equipment changes, such as circuit multiplication techniques but this must have a limit. The launch of new satellites, with improved technical performance and facilities, will impact upon the earth station requirements involving either completely new construction or modification and addition to existing equipments.

The same cannot be said of optical fibre cable, which has repeaters designed with a working life of 25 years and may not need to change any of the actual cable due to the transmission characteristics of the cable itself. What may be required is new terminal equipment to replace obsolete technology.

11.4.4 Commercial factors

A review of first order cost factors, running and upgrade costs still does not take into account what might be termed commercial factors. The transmission methods must be looked at with regard to trends in communications, particularly with regard to digital systems. The quality of communications has to be considered so that the echo characteristics become important as well as ensuring that the systems are amenable to DCME techniques. Another factor is the interference characteristics of each system, for satellites interfere both with other satellites and terrestrial systems and cannot therefore be designed in isolation but must take account of other installed systems. Optical fibre does not only produce interference in other microwave systems but is also relatively immune to man-made interference, caused by car ignition systems, etc. However the fibre system has a distinct disadvantage in that it is not immune from physical damage. One further element, related to the growth of capacity is that the geostationary arc will eventually reach a limit for satellite deployment and users will have to utilise more complex arrangements.

11.4.5 Satellite costs

For satellite communications the cost components are:

- Satellite;
- Launching costs;
- Land for earth station;
- Earth station buildings;
- Earth station antennas;
- HPAs;
- LNAs;
- GCE equipment;
- Installation costs;
- Multiplex equipment;
- Echo cancelling;
- DSI/LRE.

The costs of the above items are detailed in Table 11.4. These costs relate to an INTELSAT 6 satellite, which uses both voice and data transmissions with the applications of DSI and LRE to give an operational capacity of 55 000 circuits, with a total capacity of 123 750 circuits.

Table 11.4   Overall cost of INTELSAT 6 satellite operation

Item	Timescale (years)	Cost (£ million, 1988 prices)	Cost/Year (£ million, 1988 prices)
TDMA E/S (3 off)	15	88	2.7
Capital cost (per satellite)	10	247	24.7
Running costs	1	6.7	6.7
Total	·		34.1

The above assumes that the cost of both satellites and earth stations involves a 14% cost of capital element. The number of earth stations is determined by the cable configuration, which has three landing points.

11.4.6   Optical fibre cable costs

The elements that must be taken into account in calculating the overall cost structure are:

- Optical fibre cable costs;
- Cable installation costs;
- 140 Mbits terminal equipment.

The cable is assumed to be TAT8, which is a transatlantic cable connecting the USA, France and the UK. The cable has a basic capacity of 7560 circuits, with a maximum capacity of 34 776 circuits if LRE is used to give circuit multiplication. Detailed costs are shown in Table 11.5.

Table 11.5   Overall cost of TAT8 cable operation

Item	Timescale (years)	Cost (£ million, 1988 prices)	Cost/Year (£ million, 1988 prices)
Cable/Installation	25	678	27
Running costs	1	3.7	3.7
Total cost			30.7

The financing costs in Table 11.5 assume the same cost for the capital element of the calculation in order to compare like with like.

11.4.7   Distance sensitivity

The result of the calculations in the preceding sections is to show that when the satellite and optical fibre mode of transmission are compared on a

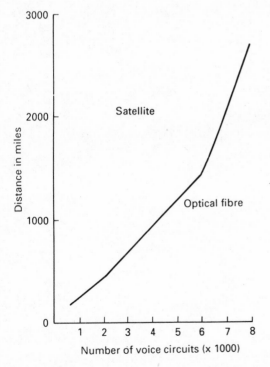

Fig. 11.4 Break-even curve for satellite or optical fibre use.

correct basis then even for long haul systems the sophisticated satellite method is the most cost-effective on a per channel basis.

The common assumption is that the satellite systems are more cost-effective at longer distances and this is certainly true, but if the first order costs, system capacity and break-even distance are taken into account then it can be seen that for trunk systems the satellite solution is very competitive down to low channel capacities and at relatively short distances as shown in Fig. 11.4.

The graph in Fig. 11.4 does not really take account of the increasing use of VSAT terminals and systems which provide a low cost satellite option and it is highly probable that the future will see the use of optical fibre as the point to point bearer terminating in satellite systems that then provide a point to multipoint capability.

Chapter 12
Future Trends and Possibilities

12.1 INTRODUCTION

The growth of traffic demand over the past twenty years has exceeded all expectations and forecasts. That pattern is likely to be repeated over the next twenty years as the world becomes more interconnected by voice and data. Various forecasts have been made one of which is shown in Fig. 12.1.

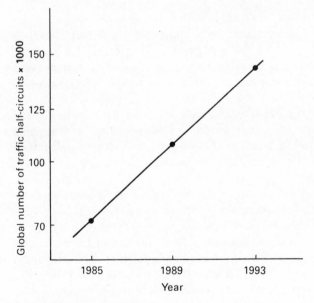

Fig. 12.1 Forecast of global traffic growth.

This traffic could be carried by satellite regardless of the development of optical fibre systems. Whether satellites can compete depends upon two factors: the ability of the satellite designer to provide more data per unit of bandwidth, for a longer period of time and at less cost, both in terms of satellite and ground station hardware; and in addition the quality of communication must be improved, with particular attention being paid to the multi-hop satellite system performance, where echo and delay are two degrading factors on any satellite link. The flexibility of the satellite system is its major advantage over other communication technologies and it must retain and further that advantage. This aspect is epitomised in the VSAT terminal type of network, which has the ability to provide a very flexible, multi-service system over a wide area and to be operational in less than twelve months.

The problems of complexity and interconnectivity involved in satellite systems will have to be solved by the use of new access methods. Indeed the whole area of access methods will have to be addressed to ensure that the equipment required for both ground station and central hub is at a lower cost compared to current costs and that it retains or improves existing performance. These considerations apply particularly to TDMA systems, whose deployment has been delayed by the system complexity and cost. The need here is to design low bit rate TDMA that can give operation at up to 2048 kbits data rate and thus bring it into the range of the small user and interface it with the burgeoning digital terrestrial systems.

The pressure on capacity will lead to further advances in bandwidth reduction to allow more voice circuits to be contained within a 64 kbits data channel. The use of ADPCM is likely to produce circuits capable of transmitting four voice circuits down a 64 kbits link. These developments will prevent satellite saturation and therefore reduce the need for more satellites, while at the same time the cost of the service can be reduced.

The satellite itself is capable of much development and later in the chapter the use of geostationary platforms will be considered, along with satellite clustering and inter-satellite links. Within the satellite the developments in technology have led to new concepts in baseband switching to provide interconnectivity; improved design of the satellite antennas to give many more and narrower beams, to allow greater flexibility and on-board processing to achieve more efficient use of the system at a higher quality.

The earth stations themselves will be subject to the same pressure of development as the satellite both in terms of flexibility and of cost reduction as well as the move to higher frequency bands of operation. The movement of technology especially in the field of RF systems and data processing means that earth stations will become smaller, more compact and able to process much more complex information. In the remainder of the chapter the various areas of satellite communications, covered in the previous chapters will be considered from the point of view of current developments and future possibilities.

12.2 SYNCHRONOUS SATELLITE DEVELOPMENTS

12.2.1 Introduction

The main elements to be considered here are:

- The current INTELSAT satellite as an indication of the direction of satellite design;
- The exploitation of Ka Band at 20 GHz and 30 GHz to increase the availability of satellite capacity;
- Geostationary platforms;
- Inter-satellite links;
- Satellite sub-system design.

Future INTELSAT satellites

The process of specifying new satellites is a complex and time consuming process and to bring a satellite from initial planning to launch can take up to 15 years. The next series of INTELSAT satellites about to be made operational are INTELSAT VI, which were fully specified in 1981 and are now under construction ready to launch over a period from 1989 to 1991. The series will have five satellites, mainly designed and constructed by Hughes Aircraft and will be of the spin-stabilised type typical of the Hughes designs. Although the design and development programme has been delayed, the development models have been tested ready for implementation.

The INTELSAT VI will probably be deployed over the Atlantic and Indian Ocean regions and are likely to fulfil the organisation's predicted capacity requirements in those regions until 1996, at which time the future capacity for the Pacific region will be supplied by the next series of INTELSAT satellites, namely INTELSAT VII. This series is interesting in that it will not follow the trend towards larger satellites with higher capacity, but will be of lower capacity and have less flexibility than INTELSAT VI as it will not use reconfigurable antenna beams. In general these satellites do not employ frontier of technology equipment, though the SS–TDMA switches and Series VI reconfigurable antennas seriously stretch any satellite designer.

The INTELSAT VI series of satellites were designed to be an improvement on earlier series in the following ways:

(a) An operational life of up to 14 years;
(b) An increase of capacity by the use of sixfold re-use at C Band;
(c) Increasing capacity by the use of more transponders;
(d) Flexibility in terms of the launch vehicle, in that the satellite can be launched on any one of three vehicles; Ariane, space shuttle or a

modified Titan III rocket. This policy is prudent especially in the light of the failure of both Ariane and the shuttle which prevented satellites being launched for a long period;

(e) Following the trend of other satellites the eirp on global and spot beams is higher, giving VSAT operational capability, with a spot beam eirp of 44.4 dBW;

(f) Use of satellite switched TDMA.

The INTELSAT VII series of satellites will be both more and less sophisticated but will be cheaper and have a life of up to 20 years.

Satellite construction

INTELSAT VI will have the form shown in Fig. 12.2 when in orbit, though during the launch period the payload antennas are housed inside the main cylindrical body of the satellite. They use an offset feed design to give high efficiency and provide four co-frequency and co-polarised zone beams, with 3.2 and 2.0 m diameter respectively, and they have a focal length of 4.2 and 2.6 m. The offset feed is an array of 146 feed horns with various groups of these feeds being excited to produce the appropriate satellite hemi and zone beam patterns. The East/West spot Ku Band antennas are also offset feed antennas, with fixed mounting between main and offset reflectors, which can be steered to illuminate any portion of the earth visible to the satellite. The antenna patterns are completed by the two global beams produced by horn antennas of different sizes for transmit and receive purposes. The satellite can be seen to have great flexibility not only in terms of its coverage capability but also in its dual frequency role and multiple access traffic possibilities.

Power requirements

The enhanced communications capability of the INTELSAT VI satellite means that the satellite power supply system must generate over 2 kW of prime power in order to sustain the 48 output amplifiers, which may be TWT or solid state in form and so provide the transmission capability as shown in Table 12.1.

The spin-stabilised satellite uses solar panels attached to the cylindrical outer drum of the payload housing; INTELSAT VI is unusual in that the effective power capacity of the system is increased by the use of a deployable solar panel. At launch, this forms a cylinder over the main body of the satellite and is extended telescopically to a distance of 38 metres beyond the satellite body once the satellite is in orbit. This system in effect doubles the power capacity at little weight penalty. The satellite is operated during an eclipse by 64 nickel hydrogen batteries.

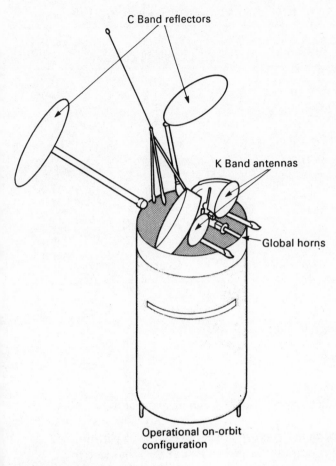

C Band reflectors

K Band antennas

Global horns

Operational on-orbit
configuration

Fig. 12.2 INTELSAT VI satellite.

12.2.2   20/30 GHz operation

Introduction

The pressure for greater transmission capacity has led to the development of
Ku Band as a satellite frequency band and in turn designers are now turning
their attention to Ka Band at 20/30 GHz, for commercial applications.

The method of exploiting these frequencies is similar to that used for Ku
Band where experimental satellites are being used to test transmission
parameters and earth station technology before full commercial services are
put in place. The advantages of Ka Band are:

- Smaller earth station antennas are required, one fifth of the size of the
  equivalent C Band systems;

Table 12.1   INTELSAT VI transmission capability

Beam	Flux density (dBW/m)	G/T (dB/K)	eirp (dBW)
W Spot	− 77.3	1.3	41.7
W Spot	− 78.0	1.7	44.7
W Spot	− 77.3	1.3	44.7
W Spot	− 78.0	1.7	47.7
E Spot	− 78.0	1.0	44.7
Hemi	− 67.1 min	− 9.4	31.0
Hemi	− 77.6	− 9.2	28.0
Hemi	− 70.1	− 9.5	28.0
Zone	− 67.1 min	− 7.0	31.0
Zone	− 67.1 min	− 2.0	31.0
Zone	− 77.6	− 2.0	28.0
Zone	− 77.6	− 7.0	28.0
Zone	− 70.1	− 3.0	28.0
Zone	− 70.1	− 7.5	28.0
Global	− 70.1	− 14.0	26.5

- Non-interference with terrestrial systems;
- Wider bandwidth available.

The disadvantages are:

- Higher up and downlink attenuation;
- Higher performance earth station components and sub-systems are required;
- Larger system design margins are needed to guarantee performance;
- Larger HPA power outputs are needed to give the required eirp.

Japanese satellite systems

There have been a number of systems put into experimental operation, with the Japanese being heavily involved in this area of research and development, with the launch of an experimental satellite CS1 in 1977, with the second series CS2 now operational having been launched in 1983. The CS2 series has provided Japan with a 20/30 GHz capability into the late-1980s. The CS2 series uses two satellites, 2a and 2b, which provide both an experimental test-bed for Ka Band systems and a practical system for use as a backup to existing services.

The characteristics of the CS2 satellites are shown in Table 12.2.

The earth station design is obviously similar to that in any satellite network, though the specific problems of Ka Band must be taken into account by the use of up path power control to overcome the fading characteristics of the transmission medium. The CS2 system uses two differ-

Table 12.2 Japanese CS2 satellite characteristics

Parameter	CS2 satellite values
Orbital position	132° E/136° E
Satellite type	Spin-stabilised
Transponders	C Band 2 off
	Ka Band 6 off
Capacity Ka Band	480/132 channels for telephony
	1 channel for TV
C Band	192–672 channels for telephony
	2 channels for TV
Capacity bandwidth	200 MHz
Access method	TDMA
Modulation method	BPSK and FM
Design life	5 years

ent sizes of earth station, characterised by their antenna size, and by whether it is used for a fixed or mobile service. The basic form is shown in Fig. 12.3.

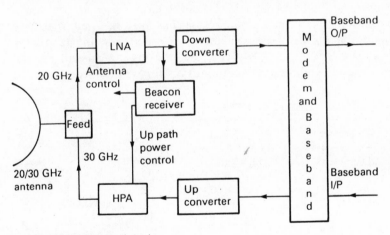

Fig. 12.3 20/30 GHz earth station.

The earth station for the fixed service uses a 5 m antenna operating in a digital mode at the primary rate. The basic earth station parameters are shown in Table 12.3.

The special requirements for the earth station relate to all areas of the equipment. These are detailed below.

Antenna
The antenna performance is detailed in Table 12.4, showing that the antenna gain is equivalent to a 27 m diameter unit at C Band. The antenna must maintain a very high surface accuracy in order to achieve this gain and the

Table 12.3 CS2 earth station parameters

E/S parameter	Parameter value
Tx frequency band (GHz)	28.4–29.0
Rx frequency band (GHz)	18.6–19.2
G/T (dB/°K)	34.5
HPA power (W)	20
Modulation	BPSK
Bit error rate	10^{-6}
Beacon frequency (GHz)	19.45

physical construction must be such that this accuracy is maintained over the whole range of environmental conditions.

The antenna is almost certainly constructed using a carbon fibre plastic material which is manufactured in a sandwich construction of fibre glass layers on top of an aluminium honeycomb, giving a lightweight, strong antenna.

At a 5 m size the antenna will be panelled, which must have a surface accuracy of better than 0.15 mm rms in order to meet the required performance. The sidelobe performance requirement has to be met and to achieve this it is likely that the feed system will use a corrugated horn. In line with the trend towards simplicity and low cost the tracking system will use the step-track method over a restricted range of X/Y axis movement.

Table 12.4 CS2 antenna parameters

Parameter	Value
Diameter	5
Material	Reinforced CFP
Type	Cassegrain
Sidelobes dBi	$32 - 25 \log \theta$ $1° < \theta < 48°$
	$- 10$ dBi $48° < \theta < 180°$
Gain Tx	61 dB
Rx	58 dB
Noise temperature °K	50 °K (45° elevation).

Low noise amplifier

The LNA must have the same general characteristics as those used at Ku and C Band, i.e. low noise temperature, high gain, wide bandwidth and high reliability. The latest type of LNA and one which is undergoing a great deal of development, uses the high electron mobility transistor, HEMT. This is capable of providing the necessary performance, within the framework of a Peltier cooled amplifier, to give a 1.7 dB noise figure. A typical Ka Band LNA performance is shown in Table 12.5.

Table 12.5 Typical Ka Band LNA performance

Parameter	Value
Noise temperature °K	140
Gain (dB)	40
Gain variation	1 dB max
Power consumption	200 W

High power amplifier

The high power amplifier requirement, at Ka Band, of a 20 W amplifier, requires the use of a travelling wave tube. This means that the form of the amplifier is as shown in chapter 4, with a GaAs FET intermediate amplifier preceding the main helix output tube. A typical set of characteristics is shown in Table 12.6.

Table 12.6 Typical Ka Band HPA performance

Parameter	Value
Frequency (GHz)	27.5–29.5
Gain (dB)	65
TWT	PPM – focused coupled cavity
TWT gain	33 dB

The overall architecture of the CS2 satellite systems is either on a TDMA or Demand Assigned TDMA according to the needs of the network. The system has found wide application in Japan for private networks, backup systems, remote island relay and mobile services. In these roles it has proved a very effective proving ground for Ka Band systems. It is likely to put Japan in the forefront of Ka Band satellite communications, both in terms of satellites and earth stations thus maintaining its commercial pre-eminence.

The Japanese space programme will continue into the future with the launch of the CS3 satellite series to retain its Ka Band capability. It will put into orbit an experimental satellite that will look at the S Band performance of multi-beam array antennas, using the ETS VI satellite.

Olympus satellite

The Olympus satellite, when first envisaged, was a very advanced concept proposing as it did the use of Ka Band and Ku Band payloads on a single platform, providing a wide variety of services such as dbs transmissions over Italy and other parts of Europe plus business data services and VSAT networks. The basic satellite characteristics are shown in Table 12.7.

The table shows that the Olympus satellite is intended to provide a high

Table 12.7   Olympus satellite characteristics

Parameter	Value
Satellite type	three-axes stabilised
Satellite weight	2420 kg
Dimensions	21 m × 1.75 m × 5.3 m
Solar panel dimensions	25.7 m
Orbital position	19.0° W
Operational life	10 years
Payload Ka Band	3 transponders 40 MHz
	700 MHz
	40 MHz
	Signal power 52 dBW
Ku Band	2 dbs/27 MHz/62 dBW
	4 telecom channels/18 MHz/44.3 dBW
TWTA	dbs 340 W
	Ka 30 W
	Ku 30 W

power dbs capability which will in fact allow very small antennas, i.e. 45 cm diameter to be used. The Ka band transponders are for experimental systems that will provide voice, data and video services throughout Europe. The 12 GHz transmissions are on five steerable spot beams, steered as a total group, with frequency re-use between spot beams. The service will operate in a TDMA access mode, providing the same facilities as the other frequencies. The major transmission problem associated with Ka Band operation is fading. Though this is also present in lower frequency band satellite systems its importance is much greater at Ka Band. There are a variety of methods to counteract fading which can cause a reduction in received satellite signal level which can be as high as 10 dB.

If up path power control is used the amplifier must be capable of varying over this range, which is a considerable disadvantage in terms of cost. The method of control varies, with the simplest method being to sense fading in both the up and down paths and keep both levels constant, though this is an unstable system when the satellite back-off is smaller than the up path dynamic range. The use of up path power control (UPPC) for multi-carrier operations (single carrier does not require UPPC) reduces the efficiency of the earth station during non-faded conditions.

The fades can be compensated for, to different degrees, by antenna diversity which must be far less cost-effective; frequency diversity which involves more complexity in the earth station; and adaptive transmission rate systems which actually reduce transmission bit rates to maintain a required link bit error rate. This latter method can only be applied if the operator can withstand this rate reduction.

The detection of fades and the response to that fade provide the system designer with problems. The fade can be detected by measuring its effect on a satellite beacon signal, that signal being detected on a separate beacon receiver. The fade can be measured by measuring the carrier/noise ratio of the signal though this has a difficulty because of the length of time taken to detect a change in BER; this will produce a lag in the system. When trying to control the power the control system must respond rapidly to the fade otherwise the earth station reaction will occur after the fade has come and gone. To overcome this the system designer must build in a margin of output power to allow for the UPPC to react.

12.2.3 Geostationary platforms

The satellite is essentially a platform on which a number of communication modules are mounted and these modules can give localised and global coverage, as required, with some flexibility in terms of re-configuration from the ground control system. The logical extension of this is to provide a platform actually built in space that could house a very large number of communication modules. This concept is now capable of realisation due to the advances in launch vehicles such as the shuttle or the USSR long term flights, in what are essentially orbiting laboratories. The major impetus for this concept has come from NASA which from about 1974 has evolved its ideas through the Orbiting Antenna Farm or OAF. This was seen as an orbiting processor/relay station and was viewed as a PABX in the sky but failed to progress further due to the high cost of such a system. In 1978 NASA instituted various studies, one of which is concerned with the geostationary platform (GP) which looked at a much broader picture than the OAF, with a view to broadening the user base of the GP to include high volume voice traffic, satellite television, remote sensing and scientific missions.

The GP is intended to carry a large number of payloads with a much longer working life than the normal satellite – in the order of 20 years, with the possibility of operational servicing using the space shuttle to replace expended fuel and batteries. The space station, initiated in 1984, pushed the process still further by setting up the idea of a manned orbiting platform in low earth orbit which could construct satellite platforms and then transfer them into a geostationary orbit.

The GP has the following advantages:

- It provides better orbit utilisation at a time when the geostationary orbit is under pressure.
- It can reduce the cost per telephone circuit or data bit.
- It can provide much greater capacity over a longer period of time.
- It could provide a country, such as Japan, with all of the facilities available via satellite using one orbital slot.

The concept of the GP has been studied by a number of agencies including INTELSAT, ESA and the Japanese Space Agency; these studies have led to a variety of ideas for a GP. The INTELSAT concept looks to an antenna farm that will provide 25 times the capacity of the conventional communications satellite. It would use a construction that not only points towards the earth for communication purposes but also for astronaut communications. The power would be obtained by photo cells and concentration systems.

In Japan the GP concept uses a long narrow bus with large antennas at each end and generates a large number of spot beams to cover the whole of the Japanese islands.

The ESA concept introduced a number of novel ideas to achieve its purpose; the first is the use of balloon inflated antennas, which will be metallised in space and can give a size up to 100 m for antennas that can be used to give spot beams for mobile systems operating at L Band. The second idea is to position the platform by the use of laser beams.

The geostationary platform is an ideal opportunity to save orbital arc and bring together the wide variety of satellite services that are now available. No doubt in the 1990s the concept of GPs will come to fruition.

12.2.4 Inter-satellite links (ISL)

The need for more capacity in satellite systems leads to the design and commissioning of more and larger satellites in the field of positional fixing systems and remote sensing where the operational coverage is global and therefore requires the use of a very large number of satellites. In essence the use of inter-satellite links is an integration of terrestrial and satellite systems in that it allows flexibility by removing the need for the signal to go in a double hop path with all the problems that that entails.

The inter-linking can be in two circumstances: first where the low earth orbit system has a large amount of information to retransmit to earth but is only in view for a limited period of time, such as remote sensing systems. This problem can be overcome provided that the information can be sent, via a laser link to two geostationary satellites that will allow information to be transmitted in real-time at any time of day. The second possibility is for transmission between two geostationary satellites which will allow an increase in effective satellite capacity. In one method the satellites are clustered close together, with links between them by laser, thus allowing the use of a number of smaller and simpler satellites at lower cost. The cluster can be added to by simpler satellites to enhance the cluster capacity thus obviating the need for the development of bigger and better satellites. The second possibility is for communication between regional satellites.

The development of inter-satellite links has to consider the system aspects as well as the methods of communicating between satellites, which can be optical, microwave or millimetre waves. Which one is used depends upon the data rate of the information and the distance between the satellites. The frequencies that can be used are already allocated as shown below:

- 22.55–23.55 GHz
- 32.00–32.00 GHz
- 54.25–58.20 GHz

The weight and size considerations point towards the use of millimetre wave communications rather than microwave though it has to be recognised that the technology in the 55 GHz band is more risky and costly. The future choice for ISL is likely to be optical communications as it not only fits into the development of terrestrial networks but also has no bandwidth restriction and is therefore future proofed in this respect.

12.2.5 Satellite sub-systems

The various sub-systems and equipment that go to make up a satellite are all undergoing continuous development to improve them in response to the pressure of customer and services demands. The following sections look at a number of these elements:

Satellite life

The life of a satellite must be made as long as possible in order to reduce the cost per channel per year and thus improve the competitiveness of satellite as opposed to optical fibre cable. The useful life of a satellite is determined by a number of factors which are not related to the communication hardware, which is designed round individual components specially designed for a space environment. The two major elements that set the effective working life of the satellite are stationkeeping and power systems.

The drift and movement of any satellite has to be compensated for by operation of the rocket engines which are powered by hydrazine. This is finite and therefore once it is exhausted the satellite cannot be kept on station. The use of liquid propellants for station keeping thrusters produces diminishing returns because if more propellant is put on board the satellite the initial weight is greater and therefore satellite costs will increase. One area of research being considered is the use of ion engines, which do not need a liquid propellant; this research has led to their use in INTELSAT VII and they will be used in other satellites such as the experimental Japanese satellite ETS VI to be launched by the National Space Development Agency of Japan. The basic form of the ion engine is shown in Fig. 12.4.

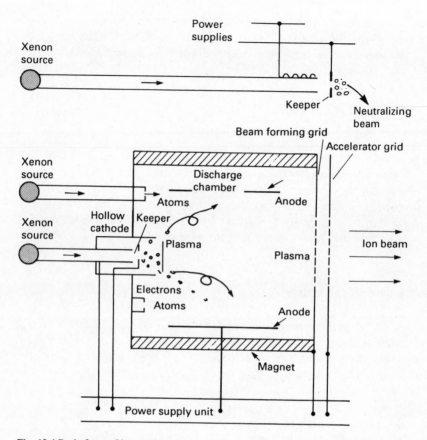

Fig. 12.4 Basic form of ion engine.

The system is made up of:

(a) Thruster units.
(b) Propellant management units that store the xenon gas and supply it to the thruster units as necessary.
(c) Power units which produce the electrical power to the heaters, anode, cathodes, accelerator grids, hollow-cathode keeper anodes and mass flow controllers. The powering of these items is carried out in a particular sequence and the unit also carries out that function.
(d) An electronic unit that controls the operation of the values between the propellant tanks and the thruster units.

The ion engine is tubular in form. The initial plasma production takes place in the lefthand cavity and is generated by injecting xenon into the gap between the main hollow cathode and a keeper anode, with a voltage being applied to both anode and cathode. The electrons in this plasma move towards the anode in the main chamber as the anode is at a higher potential, and then enter the main ion chamber. Once in the main chamber these are subjected to a magnetic field which accelerates them in a spiral path towards the anodes during which time they collide with xenon gas atoms. These collisions produce ions, a process that is multiplied by the effect of the magnetic field. The process produces positive and negative ions; the positive ions are then ejected from the chamber through an accelerator grid that produces a beam of electrons along the axis of the output port.

The problem with this is that because the positive ion beam will make the satellite itself negatively charged, the ion beam will be drawn back towards the satellite, negating its purpose. To overcome this problem the ion engine contains a second plasma producer similar to that used to provide the basic plasma source for the ion engine. The xenon atoms are fed between the energised hollow cathodes and keeper electrodes which produces a plasma adjacent to the ion beam ejected from the engine, which mixes with the ion beam to prevent ion attraction.

The principles of the ion engine, as described above, have practical problems when operational. These are:

(a) The ion production process produces flecks of material from such elements as cathodes, anodes, grids, etc. due to a sputtering effect. The flakes will deposit themselves between grids, etc., short circuiting those elements and thus reduce the ion engine performance. This problem is reduced by the use of ceramic coatings on the cathodes while heaters are given wire mesh protection.

(b) Accelerator grid corrosion due to ion sputtering which produces pitting around the grid holes.

The ion engine benefits from the fact that it discharges ions at ten times the speed of a chemical thruster discharge thus giving ten times the operational life. The ion engine is mounted on the satellite in order to generate thrust in a north or south direction and is in fact mounted at 35° relative to the axis of the satellite. The engine can produce 250 mN with exhaust velocities at 40 km/second.

Satellite power supplies and system cooling

The trend towards larger satellites, higher eirp for dbs and longer life etc. means that satellite power supplies must provide much higher power output over double the length of time for an ordinary satellite. The work horse of satellite power is the solar cell. Originally silicon was used as the basis of each cell. The solar cell is not the only method of producing power and some of the other possibilities are listed in Table 12.8.

Table 12.8 Satellite power sources

Type	Energy source
Photo-voltaic cells	Solar
Concentrators	Solar
Solar dynamic	Solar
Fuel cells	Chemical
Thermo-electric	Chemical
Fusion	Nuclear

The simplest form of power source is the photo-voltaic cell and much development has gone into this area of power generation. The latest systems use gallium arsenide cells which are more efficient and have a longer life than silicon, having an efficiency of 17% as opposed to 8% for silicon. The GaAs cell is also less susceptible to radiation damage and retains its efficiency better at elevated temperters. The GaAs cell can operate at a higher temperature and is suitable for use in another form of power supplies i.e. solar concentrators. These use a reflector, usually of Cassegrain form, which gathers the sunlight and focuses it on the solar cells thus allowing the use of smaller size arrays.

Another form of power generation uses solar heat which is concentrated by a mirror into a beam which is focused and transmitted by heat pipe into a receiver which either has an integral boiler containing a fluid that is vaporised to drive a turbine or a transport liquid which is used to convey the heat to the turbine boiler fluid. The turbine produces AC power via an alternator; the system is such that a parabolic collector would need to be eight times the effective surface area used by solar arrays.

The use of nuclear systems, either fusion or isotope will be of more use in deep space applications where the sun's rays are weaker and solar systems cannot be made large enough. Their use in earth orbit has problems as when at the end of its life, the satellite re-enters the earth's atmosphere, it brings fears of nuclear pollution.

A final possibility is the Organic Rankne Cycle (ORC) heat engine which is usually the free piston form. The system uses a fluid, such as toluene, which is vaporised to drive the turbine and alternator.

Heat control

More power implies more heat that has to be removed from the satellite sub-systems to increase operational life and prevent failure. The basic form of heat pipe is a closed system which is usually embedded in the honeycomb panels that form the satellite body.

The heat pipe has two channels, one for the fluid ammonia, the other acting as the vapour channel which carries the heat away as the gas vaporises due to the rise in temperature. Due to capilliary action the gas flows down the pipe before re-condensing and returning to the liquid channel. This system has been subject to further improvement in the following ways:

- The physical form of the pipe was changed to provide a wick effect along the pipe.
- The fluid was changed to methanol.
- A wick was inserted into the pipe.

The design is of the form shown in Fig. 12.5.

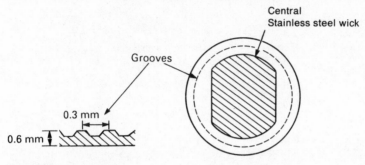

Fig. 12.5 Variable conductance heat pipe.

The pipes are efficient in carrying away the heat but it is obviously necessary for them to be mounted correctly and efficiently in order to gain maximum heat removal. One method of embedding pipes is shown in Fig. 12.6.

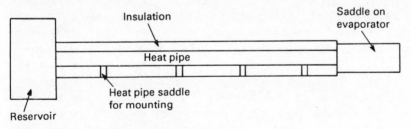

Fig. 12.6 Variable conductance heat pipe mounting.

Satellite communications technology

The services to be provided by satellite will remain the same into the foreseeable future, i.e. voice, data and TV communications with increased quality, especially in the TV sphere where HDTV systems are likely to come into widespread use. The pressure of optical fibre and other communication systems will lead designers to employ technology in the service of more cost-effective communications, implying that the satellites must become less expensive in relation to each information bit transmitted through the transponder. At the same time flexibility must increase to ensure that services can be moved as required by customer demand. The satellite must also be of such a capability that the earth station segment requires smaller and less expensive ground stations.

The above requirements mean that the satellite will require new high gain antenna designs and multiple frequency re-use to ensure that the minimum-cost solutions are used. Allied to the external improvements is the use of on-board processing; in this the aim is to reduce the size of sub-systems and the development needs can be illustrated by Fig. 12.7.

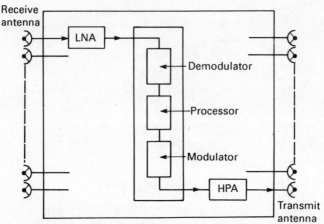

Fig. 12.7 Satellite on-board processing.

Antennas in current satellites use beam forming networks which have high losses between antenna, LNA and HPA. This loss means that larger antennas and improved amplifiers are required to achieve the required system performance. In the future the LNA will be designed with MMICs and mounted in the waveguide connecting the LNA to the receive antenna. The HPA usually uses TWTA power in the range of 5–250 W but in the future for power amplifiers up to 25 W, the use of solid state power amplifiers (SSPA) will come to predominate due to their improved efficiency and longer life associated with solid state components. The smaller HPA size allows mounting of the unit close to the transmitting antenna, with a consequent reduction in power loss.

The on-board processor either demodulates the signal before remodulation and retransmission (with an integrated circuit switch using FET switches between demodulators or modulators, linked to the appropriate output beam) or cross connects from input to output via a microwave switch, which is now being designed in a smaller and lighter form.

The availability of GaAs FET components and MMIC units means that new satellites will have increased flexibility, lower losses – which will allow smaller earth stations to be used – as well as the advantage that baseband processing removes the effect of uplink noise and distortion.

12.3 SIGNAL PROCESSING

The cost of satellite communications is directly related to the bandwidth efficiency in terms of channel throughput. The trend to digital transmission means that considerable development has gone into the various forms of modulation and the most frequently used system appears to be multilevel phase shift keying. This form of processing is intended to increase the amount of information transmitted and is complemented by coding as well as baseband processing.

The development of modulation techniques in the future will to an extent depend on the application. For instance the transmission of standard TV will probably remain within the province of FM modulation. This technique could also still apply to low capacity telephony systems that are essentially a niche application with no foreseeable development, i.e. they are unlikely to benefit from better threshold extension techniques or modifications to the form of FM modulation. There are other forms of modulation available to the future designer which have found considerable use in terrestrial systems, where they have achieved very high spectral efficiency, with values in the range of 3–4 bits per Hz. These rates are achieved by different forms of AM modulation such a multi-level QAM up to 64 QAM. In addition there are such possibilities as amplitude phase keying and non-binary PCM. The APK systems would appear to be ideal for satellite communications when considered from the point of view of bandwidth efficiency. However the fact is that the non-linearity of the satellite link means that the link degradation is worse than the PSK systems due to the inherent amplitude changes needed in the modulation process. The NBCM system is in fact a mixture of coding, as in the normal PCM system, and the modulation element is in the variation in the output levels of the PCM signal and the insertion of a carrier. One area of investigation, triggered by the development of mobile systems, is minimum shift keying (MSK) modulation. This has a compact power spectrum and a constant envelope signal, giving equal performance to that of BSPK, but in its basic form has poor performance under fading conditions; this can be improved by the use of differential detection.

The modulation system produces an improved system performance but involves increased modem complexity. Considerable improvement of information throughput can be achieved by baseband processing. This processing is already in place and corresponds with the growth of digital communications. The definition of speech processing systems is carried out by the ITU who began this process in the early 1970s with the recommending of standards for PCM systems which involved analogue processing. These early recommendations are important because they set the size of the basic speech circuit at 64 (56) kbits and from that the basic structure of the first level of the digital hierarchy. The further development in the 1980s took two direct-ions, one the reduction of speech information to allow two speech channels to be accommodated in a single digital channel – a result achieved by ADPCM using adaptive quantisation and prediction as decribed in chapter 3. The pressure on channel costs is encouraging designers to investigate speech compression techniques that will allow a single voice channel to be transmitted into as little as 2.4 kbits. The standardisation of specification on speech coding is currently considering 16 kbit techniques; such systems find use in:

- Low C/N digital satellite links;
- DCME systems for satellite links;
- Digital mobile satellite systems.

The major design effort is in techniques known as hybrid coding, in particular a modified form of linear predictive coding (LPC), known as multi-pulse LPC, which uses discrete pulse excitation for the LPC rather than white noise giving a higher quality speech. This form of implemen-tation, when used with pitch prediction, provides good speech quality with processing delay of less than 2 ms.

The use of coding provides improvement in the quality of the digital transmission links. This improvement depends upon the form of code and in general for digital satellite communications this has meant coding used for error correction purposes. This use of coding is not simply to achieve better performance but also to allow better use of access schemes or to actually achieve a random multiple access system. The development of codes presents increasing complexity to the designer and it is the availability of increasingly complex VLSI that allows the use of newer more complex codes. It is certain that future satellite systems will have higher data rates and improved error correction performance. The code system used depends on the structure of the transmission data and for higher data rates the commonly used Viterbi decoders present serious difficulties for the cost-conscious designer in that they require very complex circuitry for current designs. This is because they examine all the paths at all trellis levels and would require even more complex circuit realisations as the data rate increases. In future designs there may well be a move away from current codes towards the use of punctured convolutional codes which can accommodate higher data rates without an increased system complexity as they operate by periodically removing bits from the encoder output. This has the effect of increasing the code rate, a difficulty that can be minimised by the right choice of low rate code from which the punctured code is derived. The performance of the punctured code can be judged against the Viterbi method by considering the coding gain as in Table 12.9.

Table 12.9 Coding gain for various types of coding

Coding	Coding gain (dB)	Coding rate
Viterbi	5.2	1/2
Punctured	4.9	2/3
Punctured	4.6	3/4

There is much development that can be done on coding with a view to improving coding gain and at the same time reducing the complexity of the modem. The use of digital processing for digital modulation and coding could bring the two processes together, giving rise to the concept of the MODEC which is the combination of the modem and codec and has already shown performance improvement for 150 Mbits transmissions both for bandwidth utilisation and power required.

The increasing use of star systems in business networks and the general development of satellite systems is bound to focus the designer on the area of methods of access, in fact there is likely to be a mixing of methods. There are four basic forms of access, frequency, time, code and space, which can be combined as required. The future problem is not one of improving the existing multiplex and access methods but is more to do with producing a multi-station network that is efficient, flexible and can be adjusted to the demands of the customer. These requirements are often incompatible especially efficiency and flexibility. The current cost-effective method of allowing the user to access the satellite is some form of demand assigned format. In certain circumstances, however, when bandwidth is not a problem, then spread spectrum methods are very effective.

A system that would be very effective and may be developed in the future would effectively be a combination of a time frequency and a spread spectrum system which would effectively be a random multiple access (RMA). This system would thus have the inherent advantages of spread spectrum, in that it would be resistant to interference, while at the same time get away from the problems that pre-assigned or demand assigned systems have when dealing with a set of messages that are unpredictable, random in time and of varying length. For other access systems there is a limitation on bandwidth availability, as in FDMA, or the system is limited by the allocation of each individual time slot to a particular message, with a need to go to a higher bit rate in order to increase the capacity of the system as in TDMA. The RMA system would also have to be flexible enough to allow any terminal to communicate with any terminal without the need to check that a transmission channel is available, at the same time having simple control systems and reasonably simple equipment configurations.

The modern VSAT systems use some form of ALOHA method to control the flow of messages within the network but even there the efficiency of the method is never much better than 50% whereas the RMA method, due to its spread spectrum characteristics, would allow overlap of signalling messages and therefore is not affected by message collisions and thus can allow 100% channel utilisation.

12.4 EARTH STATION TECHNOLOGY

The minimum antenna size is limited by system considerations with key elements being better efficiency and sidelobe performance; this is done by increased efficiency and advanced antenna design techniques. The development of antennas has concentrated on off-set dual reflector systems, with a variety of reflector shapes and due consideration being given to low-cost manufacturing techniques. A system using a sector-shaped Gregorian dual reflector can give a 75% efficiency, which is better than the current designs, a − 45 dB cross-polarisation characteristic and can be made simply from stretch-formed panels and needs only a simple feed. The antenna performance is optimised by shaping of the main and off-set reflectors.

The trend in earth stations is towards smaller size and improved performance. These trends have been reinforced by the need to compete on price, performance and flexibility with other communication mediums. The major trend is towards smaller earth stations as for example in VSAT systems which are not only small in size but are co-sited with the terrestrial equipment.

The low cost of these earth stations means other costs such as installation, maintenance and manning become very important in considering the total life cost of the system. These considerations mean that although it may increase the cost of the earth station it is necessary to ensure that each module can provide remote alarms and information to a central control system. The reliability of earth station equipment is already extremely high and the increasing use of digital transmission techniques, together with the ability to use VLSI integrated circuits for baseband signal processing and MMIC modules in the RF systems, will increase the earth station availability even further. Maintenance of equipment is labour intensive and station-keeping even more so. It is therefore very important to have unmanned earth stations with remote control systems and high system availability, with redundancy systems for the key elements such as LNAs and HPAs.

The low noise amplifier designs that will be considered in the future will utilise HEMT transistors that can provide input noise temperatures of 35 °K with 12 dB gain at 12 GHz. These transistors have tremendous potential not only for low noise performance at Ku Band but also for Ka Band and above. For example it is already possible to obtain a 4 dB noise figure at 60 GHz, over a bandwidth of 3.5 GHz.

High power amplifiers are the key cost and performance element in the earth station system, with development of solid state amplifiers being a very important activity for the future. The major problems are the amount of power output available from the solid state power amplifiers (SSPA) and the cost per watt of such amplifiers. It should be said that the problems are diminished by the changes in total system design with the use of higher power satellites, thus requiring lower earth station eirp. The basic elements in the SSPA are the power GaAs FET and over the past ten years the performance of such devices has vastly improved due to better manufacturing techniques, built-in heat sink design and MMIC technology. The current developments are looking to FET and HBT (hetero-junction bi-polar transistors) and HEMTs to provide improved performance beyond current levels. Current device developments rely on the further improvement of manufacturing technology and system design to achieve better performance. Uniformity of devices is achieved by ion implantation, uniformity being necessary as the total device is simply a large number of GaAs FET cells paralleled together. Device life is very important and this is dependent upon the channel temperature which is a function of thermal resistance and dissipation in the semi-conductor device. This is reduced by reducing the thickness of the wafer and putting a gold plated layer beneath the wafer. All these techniques lead to lower impedance and it is necessary to insert matching devices into the FET to reduce overall amplifier costs leading to simplicity of construction and testing. These techniques will be further developed for the future but these are only incremental and to achieve a step in power output at higher frequencies new devices must be designed.

The HBT is an improvement on silicon-based devices at higher frequencies because base resistance and emitter capacitance can be reduced and current devices can give 1 W output at 5 GHz, with a better power/emitter length capability than GaAs FETs. InP (indium phosphorus) MISFETs have characteristics that point towards considerably better performance than the standard GaAs FETs because of their lower thermal resistivity and higher breakdown voltages.

The problems to be overcome with these devices are in the manufacturing technique used to form the oxide layer and its insulation properties (both in terms of performance and stability of the process).

In practical amplifiers the power output is not the only parameter that has to be considered; linearity is very important in order to minimise third order intermodulation products. In the GaAs FET, non-linearity is a function of FET construction at low input levels and the current and voltage limits at high input levels. Considerable improvements are achieved by the use of special profiling of the FET doping to allow the use of much higher supply voltages and thus lower third order intermodulation distortion.

Amplifier development has not only concentrated on FET devices but also on practical methods of distortion compensation. The classical method of achieving this is to apply distortion to the signal prior to the HPA input; this distortion attempts to be the inverse of the HPA characteristic and thus cancel out the basic unit distortion. It is possible by the use of pre-distortion and FET design to increase the third order intermodulation point by 5 dB. For higher power requirements the future lies with TWT amplifiers which can deliver higher power with wider bandwidths for both satellites and earth station systems. Their performance can be improved in both life and size terms by closer quality control and tungsten cathode impregnation by barium.

One other area of component development is in VLSI technology and this requirement is exemplified in the use of a transputer to provide the processing required in a GPS receiver, the outline of which is shown in Fig. 12.8.

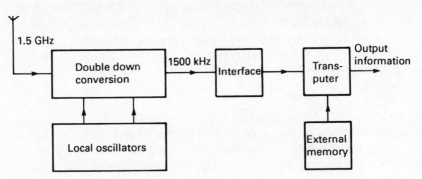

Fig. 12.8 GPS receiver using transputer.

The receiver takes in a signal at L Band, converts it down to 1500 kHz, before application to the transputer to recover the positioning information. This development allows the production of very small, hand-held receivers at low cost for quantity production.

12.5 TELEPHONY SYSTEMS

The development of telephony is synonymous with the development of INTELSAT and in chapter 5 this history is set out up to the present day; it is perhaps also the future for telephony. Much has been made of optical fibre and its threat to satellites but already it is obvious that the fibre systems require a bypass communication link in order to protect against cable failure. One difficulty in the future is to determine the rate of growth of overall services in order to plan not only voice but also TV services on a global basis because that rate is variable from region to region. In Table 12.10 an average of basic forecasts is shown for data traffic requirements.

Table 12.10   Forecast of global data traffic growth

Year	Data traffic (k Mbits
1989	4.8
1990	7.0
1994	17.0
1998	28.0
2000	33.0

It is also interesting to consider the US market which can be divided into voice, data and video etc. and this is shown in Table 12.11.

Table 12.11   US traffic growth forecasts

Service	Number of channels (Yr 2000)
Voice	1.25 million
Video	600
Teleconferencing	4250
Data (Mbits)	19 000

It can be seen that the forecast shows that there will be a slowing down of communication requirements to the year 2000 though the breakdown of figures shows that the mix of data and voice rate of growth changes from 1995 to 2000.

12.6   MOBILE SATELLITE SYSTEMS

This area of satellite communications now represents not only one of the most exciting sectors but also the one with the greatest potential in terms of growth, as it now embraces sea, land and air mobile systems, which in themselves divide into two areas of use; communications and positioning systems.

The International Civil Aviation Organization (ICAO) oversees the global requirement of aeronautical systems which covers not only communications but also positioning and surveillance. The ICAO has reviewed the future requirements through the use of the Future Air Navigation Systems Committee (FANS). This committee has strongly recommended that the future of aeronautical communications and surveillance systems lies with satellites and this should be an urgent task to complete. There are a number of systems currently being put into operation which would be suitable and it would be unwise to restrict the future to a single implementation. Indeed, as was seen in chapter 6, there are great cost and performance benefits to be gained by co-operation rather than competition, though currently the US-designed GPS system is the most advanced. The GPS type system is seen as being suitable for all types of conditions from low density telephony to aircraft on intercon-tinental flights to high density telephony over continental land masses and airline terminals, with satellites seen as replacing the current guidance systems such as LORAN C. The other area of use being considered, over and above communications and positioning, is that of ground surveillance systems that could be used in air traffic control systems to build up a three-dimensional picture of the aircraft and its position. This would allow air traffic control procedures to know the aircraft position more accurately and so allow more aircraft to be in operation in the already crowded air space but with increased safety. The FANS committee has also done cost benefit analysis for the use of satellites, the results being shown in Table 12.12.

Table 12.12   Aeronautical mobile system annual costs

Element of cost	Element value (£ million, 1988 prices)
Satellites	630
Aircraft equipment	2500
Ground equipment	140
Miscellaneous	500
Total capital cost	3770
Annual depreciation	220
Yearly operation/maintenance cost	375
Total annual cost	595

The costs shown in Table 12.12 can be set against the perceived benefits of the use of satellites, these benefits being lower operating costs and higher efficiency in the use of aircraft. These are set out in Table 12.13.

Table 12.13   Aeronautical mobile system cost benefits

Yearly benefits	Benefit value (£m)
Lower costs	630
Higher efficiency	2860
Total benefits	3490

The satellite costs assume a ten-year life cycle, though other costs are taken over a 20-year period. It can be seen that the high level of benefit from the use of satellites would allow the cost of setting up the system to be recovered in less than two years.

From a telecommunications point of view the aeronautical market mirrors the maritime, in that the first users of telephony and data services will be in the large aircraft using major routes. In this sphere there is a predicted growth of 33% to the year 2000 from a base of 9000 aircraft. Of this traffic 6% is transatlantic and now the various systems are under way is likely to grow very quickly.

The maritime system that is likely to dominate over the foreseeable future is INMARSAT, whose major element is maritime communications with a growth of total ship market from 5000 in 1986 to a predicted size of 17 000 in 1995; this market is sub-divided into large and small ship segments which are served by Standard A and Standard C ship earth stations respectively. The future trend will be to replace the Standard A with a new Standard B system which will allow more ships to be served by INMARSAT, due to better utilisation of bandwidth and satellite power. The Standard C system is now undergoing testing and will be used on very small craft into the future.

One of the major innovations, due to be implemented by 1991, is the Emergency Position Indicating Radio Beacon (EPIRB), which is part of the Global Maritime Distress and Safety System (GMDSS). The EPIRB units will work at L Band and either be free floating or automatically activated if a ship should actually sink. The EPIRB has undergone trials to prove its performance and reliability and has been accepted by the International Maritime Organization as the equivalent of the 406 MHz system.

The land mobile sector of the INMARSAT system will use a Standard C terminal and has to compete with the growth of the cellular mobile radio systems, which are growing in parallel to satellite systems. Cellular radio is, however, not perfect and has both technical and operational limitations and the use of satellites can provide a more uniform coverage of continental USA and Europe.

The problems for the mobile satellite sector in the future are not so much to do with the detail of equipment design but how to increase the amount of frequency spectrum, not only to allow the service to expand but also to provide an increased number of services. There are a variety of ways in which this can be done by:

New frequency allocations

The frequency spectrum allocated to mobile systems, shown in chapter 6, can be seen to be limited and the additional land and aeronautical services that will be developed over the next twenty years will only add to the pressure on that spectrum. There will be some additional bandwidth made available to the user in the short term from 1545–1548 MHz and in 1646.5–

1649.5 MHz but it is likely that a **WARC** Conference will be needed, in the early 1990s, to solve the allocation problem.

Frequency re-use

Frequency re-use is not a new technique but for future mobile satellite systems the system can be very complex. This is because the use of different polarisations not only adds to the complexity needed to switch polarisers but also restricts the choice of mobile ground station antennas, as some antennas cannot operate on different polarisations or meet the required operational performance. If frequency re-use is achieved by using different spot beams then new technology must be found to provide low-cost ways of connecting from input to output beams.

On-board processing

The trend to on-board processing can be of great value to mobile services in that it can achieve flexibility of bandwidth allocation thus increasing the effective bandwidth available.

The most complicated form of mobile system facility is the global positioning system. This complexity brings problems of cost due to the number of satellites involved. This can be approached in two ways:

(a) The spread of the costs of the systems between more than one operating agency;
(b) Design new systems.

The first option is perhaps already under way not only in the GPS/GLON-NAS case but also as ESA and the USA come closer together in other areas such as the space platform. The second option is already under way and is being considered by a number of agencies, the proposed system is known as STARSAT which is a concept put forward by a USA company called Starfind. The system is revolutionary in that overall global coverage can be achieved by the use of five inflatable satellites in geostationary orbit, with far-reaching implications on satellite, launch and operating costs.

The unique feature of the system is that the satellites are given axis movement at 12 rpm and yaw movement at 1 rpm. The satellite concept is shown in Fig. 12.9 and consists of a standard cylindrical form of satellite mounted above a large 36-spoked receive antenna which radiates out from a central hub. The base faces the earth and that earth facing element contains the downlink receiver and transmit systems for satellite control from the ground control. Above the satellite is a sun facing solar array to power the satellite sub-systems.

The ground based part of the system consists of individual transceivers and a ground control centre which carries out the complex signal processing

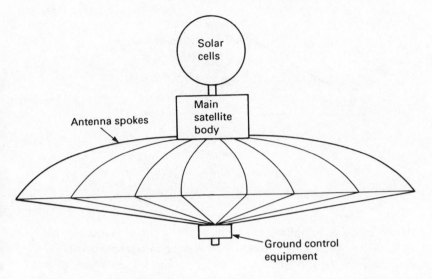

Fig. 12.9 STARSAT satellite system.

needed to pinpoint the physical location of the transceivers. It is derived from the basic unprocessed data sent from the satellite and the mapping databases held in a central computer.

The basic principles of operation rely on three elements:

- Satellite spin;
- Satellite yaw;
- Satellite antenna spoke spacing.

Each of the 32 antenna spokes 'sees' a specific portion of the earth's surface, a coverage that is determined by the spinning of the satellite. Each spoke creates a rectangular receive pattern at the earth's surface. The leading edge of the rectangle has to cross the transmitter on the ground and signals can be received as long as the transmitter remains within the rectangle, but are lost once the back edge of the rectangle passes over the transmitter. The transmitter position is determined because during the period when the transmitter is in the receive rectangle the yaw motion of the antenna, at 1 rpm, moves the apparent antenna position putting the line of loss and the line of acquisition at an angle and across one another. The transmitter is situated at that crossing point, which defines the transmitter position on a virtual axis of rotation, which is the line where loss and acquisition lines are extended into space to form planes which cross at the VAR. This process is repeated on each antenna spoke, the two spokes produce two VARs as shown in Fig. 12.10, which are two edges of a tetrahedron and form a third edge at the satellite.

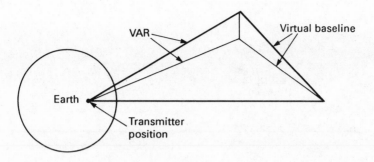

Fig. 12.10 STARSAT geometry.

If we then consider a mid-plane way between the acquisition and loss planes then we have mid-plane VARs which, when extended into space, meet at the slant line, the fourth line of the tetrahedron. If a line is then drawn through the focal axis from where the VARs meet the antenna until it meets the slant line then we have two virtual baselines that complete the tetrahedron. In order to achieve an accuracy of better than four metres the length of the virtual baselines must be at least 1000 km, a distance achieved by the use of the proposed number of antennas on a geostationary satellite.

The satellites will be made of Kevlar and inflated in space by the use of nitrogen. Once inflated the Kevlar will cure in space. The antennas are adjusted and hardened off once they have been set up. The satellite is so light that it can be launched on a simple private rocket system at a tenth of the cost of normal geostationary satellites. It also has a very low cost transceiver which is made even simpler by the fact that each receiver is identified by the irregular spacing of the antenna spokes and therefore does not need special identification decoding circuitry.

12.7 BUSINESS SYSTEMS

The technology to provide the business community with reliable communications, particularly for the transport of data is already in place. These communications have to be reliable, flexible and comparable in cost to other forms of carrier. The key to the development of business communications is not technology but regulation and the on-going effects of de-regulation, as well as the increasingly broad range of services now available as an integrated package.

In future, businesses will be looking first for data communications, at up to a primary rate, secondly for voice communication and thirdly for video and image transfer capability. The first and third requirements may well come at the same time as video codecs are now available that allow full motion video using 384 kbits and videophone using 64 kbits as a carrier. The video-conferencing facility with full motion is a growing satellite service for remote meetings, training and transfer of information but it is likely that the de-regulation will make it a part of a company's private network which integrates all three services together.

For satellite systems the major problem in the take-up of voice services is the customers' perception of the effect of delay, due to satellite path delay. This perception is now out of date but the users' perception has not caught up with the technology and it is a matter of marketing of the product to convince prospective users that they will have no problems.

12.8   VERY SMALL APERTURE TERMINALS

The future will see little difference between business and VSAT terminals as the convergence of technology brings the two systems together, at least at low to medium data rates. This convergence is not simply one of hardware but also of digital bit rate capability. VSATs will converge from a low bit rate point of view and conveniently meet at the basic ISDN service which is based on the use of a 64 kbits data channel, allowing the user to carry two duplex 64 kbit bearer (B) channels and a 16 kbit delta (D) channel. This bearer will not only allow the user the use of a switched circuit for standard ISDN, which will support real-time communications and the transfer of bulk computer data, but will also provide a wide variety of other services including telephony. The latter will occupy 7 kHz, for packet services that are more suitable for such teleservices as telex, telemetry, teleshopping, etc. and other teleservices such as videotex, viewphone and audiography which will use 64 kbits. The move to ISDN will present VSAT networks with more problems than simply controlling the passing of messages between terminals. It will need to process the signalling information contained in the total signal; this will be common channel signalling which will be the means of setting up the links, supervising and then terminating them. In this respect the CCITT signalling system No 7 will be improved in order to support the ISDN service.

The definition of the ISDN service for terrestrial systems has taken a long time to come to fruition and has been a very evolutionary process. During that time CCITT have also detailed the characteristics of the 64 kbits service in Recommendations G106 and G821 which define the hypothetical reference connections for a satellite link. This HDRP is set out in CCITT Rec 521–1 which proposes a one-hop system that can also involve an inter-satellite link and also support various access systems and the use of DSI. The hypothetical link is defined in terms of bit error rate and service interruption.

The packet switching facility service on ISDN will allow the VSAT network to replace terrestrial LANs, providing in fact a wide area network which uses the hub station to up-link packages to the satellite which then distributes them down to each VSAT remote terminal. This development fits in very well with the future use of on-board satellite processing which will control the link performance as well as acting as the master control for the VSAT network, thus reducing the cost of the central hub station.

The further development of the VSAT network as a provider of ISDN services will run in parallel with the move towards wideband ISDN and this would be to B–ISDN. However at this time it is unlikely that VSATs will be able to extend themselves to cover all the proposed services to be supported on B–ISDN, these being:

- Broadband video conferencing;
- Very high speed digital transmission;
- High speed fax;
- High definition television.

These services are set out in CCITT Rec I 130 which attempts to define what is meant by the quality of such services, at the same time endeavouring to restrict the number of channel definitions in order to prevent a proliferation of new standards.

12.9   SATELLITE TELEVISION

Satellite television has experienced major growth since 1983 and it is expected that this area of satellite use will continue to grow. The main question to be asked is how quickly it will grow. The other direction is a technical one in which the dividing lines between satellite distribution of television and direct broadcasting by satellite, dbs, are beginning to blur; in the future we may see these two areas of satellite use using similar sized antennas. The speed of growth is to some extent a matter of technology change but, as experience has shown, unless it is possible to reduce the individual receiver costs to a level that make receivers available to all economic groups, the major obstacle in the way of the rapid spread of dbs is the programming offered. This is exemplified in the Japanese experience by the NHK, which is the Japanese public broadcasting service and essentially a non-profit making organisation funded by a licence fee paid by the user. NHK has two satellites in orbit, with a third to be launched in 1990; a full scale dbs service is provided on a 24-hour basis and this has now grown from 140,000 subscribers in 1987 to 1.5 million in 1989, with a predicted growth to three million in 1990. The significance of this example is that the success of NHK is not due to profit but to targeted programming.

The trend to smaller size is shown in the development plans of EUTEL-SAT with the EUTELSAT II series of satellites. For the current series the

TVRO reception uses 1.2 m to 1.8 m diameter antennas but for the Series II satellites the antenna requirements will approach a true dbs system even though the satellite power will only be 50 dBW at the beam centre, as shown in Fig. 12.11.

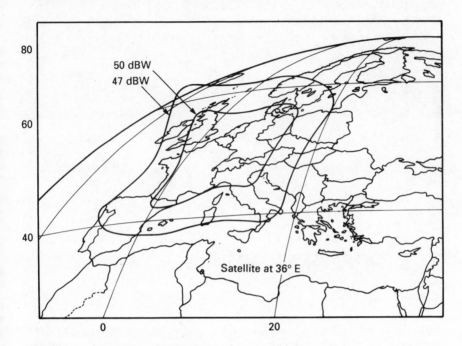

Fig. 12.11 EUTELSAT II dBs footprint.

The above satellite characteristic will allow the TVRO earth station shown in Table 12.14 to be used:

Table 12.14   EUTELSAT II TVRO parameters

Antenna diameter (m)	0.6
Antenna efficiency	65%
LNA Noise Temperature (°K)	70
C/N Ratio (dB/°K)	13 (clear sky conditions)
S/N Ratio (dB)	52
Picture quality	Grade 4 (PAL)
Climatic degradation	− 2 dB for 99% of worst month

This scenario is based on the assumption that there will be a steady improvement in LNA technology, a not unwarranted assumption if we consider Fig. 12.12 which shows the progress of HEMT technology in the development of LNAs and its expected progress to 1992.

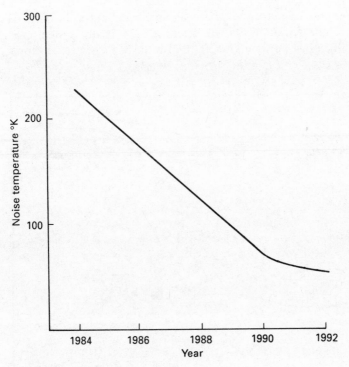

Fig. 12.12 HEMT noise temperature.

This improvement in technology in the future will affect not only the LNA element of the TVRO system but also the antenna design. Because of the very large potential size of this market it is likely that the receiver front ends will utilise MMICs, giving small size and wide bandwidth.

There are many flat alternatives to the paraboloid antenna and in Table 12.15 the current performance status of the individual systems are detailed.

Table 12.15   Performance of various flat plate antennas

Antenna type	Size	Gain (dBi)	Efficiency (%)
Microstrip wide slot	–	25	–
Microstrip line slot	–	34	–
Comb-line	–	33	20
Crank type microstrip	640 × 430 mm	34	60
Strip dipole	354 × 272 mm	29	–
Square patch	680 × 360 mm	34	60
Folded dipole	–	37	50
Rectangular slot	420 × 600 mm	34	65
Circular patch	556 × 476 mm	34	–
Radial line slot	600 mm dia	36	76

The table shows that the dbs antenna can have high gain and small size at reasonable efficiencies. In the future the most promising area of development would seem to be the rectangular slot array and the radial line slot array, which have very high efficiencies and gain for a reasonable size. The radial line slot array is one realisation of the family of slotted waveguide antennas whose basic form is shown in Fig. 12.13. The radial line waveguide is made up of three plates, the top one of which has radiating slots. The antenna is fed by a coaxial cable which is transformed to waveguide within the structure; an inward travelling wave excites the pairs of slots and grating lobes are suppressed by the slow wave structure as shown. The other important characteristic of the waveguide antenna is that the beam can be tilted. This not only improves performance but also allows the pointing direction to be altered and so give the user the ability to see more than one satellite, without the need for cumbersome and expensive actuators and antenna control systems.

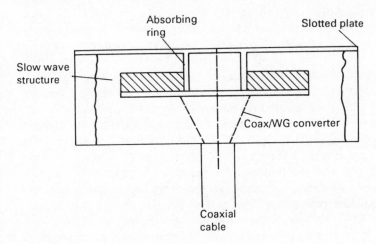

Fig. 12.13 Radial line slot array antenna.

12.10 REMOTE SENSING BY SATELLITE

The predicted retrenchment of satellite into niche user areas, due to the spread of optical fibre cable systems, would undoubtedly leave remote sensing as one of those niche markets due to the increasing need to monitor the earth and atmosphere on both commercial and environmental grounds.

The most common use of remote sensing satellites is weather forecasting where the information obtained is used in a variety of ways to assist weather forecasters. In the 1990s short-term forecasting, up to six hours ahead, will no longer be a subjective process but will be based on a short term model derived from satellite data, that data being derived from such satellites as ERS1 and NOAA every three hours to provide a much improved service. In the medium-term, for forecast between six and 36 hours ahead, the satellite data will supplement the land based information to produce an overall improvement in the total information. In addition to better information from the satellite, there will be a great improvement in the interpretation of the data sent back from the satellite. The ERS1 satellite will provide improved wind sensing equipment and will be part of a number of global and atmospheric experiments that will give details of the cloud, haze and total water vapour content of the atmosphere.

One of the other areas of remote sensing is wave investigation, which is most useful for assisting both deep and coastal water shipping in their routeing and control. There has been some work carried out in this sphere, first with Geosat, which used a radar altimeter, and secondly, with the NASA satellite Seasat, launched in 1978. It was operational for four months and then the SAR results were subjected to four years of analysis. The 1990s wave monitoring will be carried out by Europe's ERS1 and 2, TOPEX and Radarsat; the measurements of wave size and formation has been found to be very difficult even for ocean based instrumentation. Satellite information will be invaluable in this area of measurement because it will give global coverage and thus provide more inter-connected information. This inter-connected information will allow the researcher to see which waves are primary and which are dominated by swell. The SAR payload will only provide intermittent information, due to the amount of data generated and the difficulty in relaying it back to the processing centres. It is therefore in this area that much work will have to be done in order to obtain a coherent picture of wave motion over a shorter time span than is currently being done.

The third area of remote sensing is earth imaging and it is to be expected that future satellite payloads will be able to provide more detailed information. On ERS1 for instance, the sea surface temperature will be measured to within 0.5 °C, even with 80% cloud cover. The earth's surface temperature will be measured to 0.1 °C, with 1 km resolution and over a 500 km swathe. These measurements will be made using infra-red and microwave radiometers with the information being stored on on-board recorders that have a 6.5 Gbit capacity which can be read out at a rate of 15 Mbits/s.

12.11 SATELLITE AND OPTICAL FIBRE

The major advance of optical fibre in the 1990s will be into the terrestrial local loop of telephony systems, which will to some extent erode the advantage that satellites have in providing point to multi-point communications. However, the coming of the ISDN service will also act to the disadvantage of optical fibre in that it will mean it cannot provide an integrated ISDN service. What is more, with the improvement of digital echo cancellers delay will effectively be removed from the satellite system. It is unlikely that the position of satellites will be usurped by optical fibre and they will be used in parallel to provide a fail-free service to the customer. In addition the penetration of fibre on a continental or worldwide basis will almost certainly take place on a slow and sporadic basis making it rather late to be of use in the spread of ISDN across Europe, for instance. It is more likely that the satellite and optical fibre systems will integrate into a total network provision both on a primary and a back-up level.

12.12 SATELLITE COVERAGE

No attention has been paid to the use of elliptical satellite orbits, though their use is becoming a viable alternative to the geostationry orbit which is becoming increasingly crowded. This pressure for elliptical orbits is again driven by the needs of satellite users who are using them for such purposes as navigation, weather forecasting, remote sensing and maritime communications – which cannot be covered by the geostationary systems. This need means that the research into the alternative satellite constellations will increasingly have to consider the most efficient methods of providing global coverage. These studies are difficult to compare because of the variability of parameters that can be built into the design of satellite networks. For instance in the 1970s the number was considered to be five, if one assumed the satellites to be in a circular orbit, however if the orbits are made elliptical then the globe can be covered by the use of only four satellites.

Appendix 1
Satellites launched from 1976–87

Year	Organisation	Orbit
1976	MARISAT F2	72.5° E
	Palapa 1	82.0° E
	Raduga 2	86.2° E
	Ekran 15	99.2° E
	Comstar 2	283.9° E
1977	Ekran 2	88.0° E
	GOES 2	246.8° E
	INTELSAT IVA F4	338.5° E
	Himawari 2	114.0° E
1978	OTS 2	3.6° E
	Gorizont 1	24.5° E
	ETS2 (Japan)	115.3° E
	INTELSAT IVA F3	177.0° E
	YURI (Japan)	325.0° E
1979	Gorizont 3	40.5° E
	Raduga 5	73.0° E
	Ekran 4	81.0° E
	Gorizont 2	90.4° E
1980	Raduga 6	46.5° E
	Ekran 6	63.0° E
	Raduga 7	229.1° E
	SBS F3	260.9° E
	GOES 4	316.8° E
	Raduga 7	335.7° E
	Gorizont 4	348.8° E
	INTELSAT V F2	359.0° E

1981	METEOSAT 2	1.4° E
	Ekran 7	49.4° E
	Raduga 10	81.6° E
	Ekran 12	99.0° E
	Raduga 8	117.6° E
	INTELSAT V F1	174.0° E
	ESA/INMARSAT	178.0° E
	RCA/Satcom	229.0° E
	GOES 5	251.7° E
	SBS2	263.0° E
	GOES 7	285.0° E
	INTELSAT V F3	307.0° E
1982	Raduga 11	34.4° E
	INTELSAT V F5	62.9° E
	Gorizont 5	95.8° E
	Gorizont 6	139.7° E
	Satcom 5 F5	217.1° E
	Westar 5	237.5° E
	Anik C3	242.5° E
	Anik Di	255.6° E
	Westar 4	261.2° E
	SBS 3	265.0° E
	RCA Satcom 4	278.0° E
	INTELSAT VA F4	325.5° E
1983	ECS 1	13.3° E
	Ekran 2	47.5° E
	INTELSAT V F7	65.9° E
	INSAT 1B	73.9° E
	Raduga 12	76.0° E
	Palapa B1	108.1° E
	Sakura 2A CS–3A	132.0° E
	Sakura 2B CS–2B	135.9° E
	Gorizont 8	139.7° E
	Raduga 13	155.0° E
	RCA Satcom 6	221.0° E
	Galaxy 1	226.0° E
	Anik C2	250.0° E
	ATT Telstar 301	264.1° E
	Comstar 4	284.0° E
	Galaxy 2	286.1° E
	Ekran 2	290.8° E
	Gorizont 8	335.0° E
	INTELSAT V F6	341.5° E
	Gorizont 7	344.0° E

1984	EUTELSAT 2–7E	3.6° E
	Raduga 14	74.7° E
	Ekran 13	76.0° E
	Gorizont 10	79.6° E
	Yuri 2A	110.0° E
	Raduga 15	128.2° E
	Himawari 3 GM 3	140.2° E
	Syncom IV–2	180.8° E
	Anik D2	249.6° E
	Galaxy 3	266.5° E
	SBS 4	268.9° E
	ATT Telstar 302	274.0° E
	MARECS B2	334.0° E
	Telecom 1–A	352.0° E
1985	ARABSAT 1–A	19.1° E
	ARABSAT 1–B	26.0° E
	Raduga 17	35.5° E
	Gorizont 11	52.7° E
	INTELSAT VA F–12	61.7° E
	Ekran 14	89.0° E
	AUSSAT 2	156.0° E
	AUSSAT 1	160.0° E
	AUSSAT 3	164.0° E
	Raduga 16	169.6° E
	INTELSAT V F–8	180.0° E
	Raduga 16	190.6° E
	Contel ASC 1	232.2° E
	ATT Telstar 303	235.0° E
	Morelos 2	243.5° E
	Morelos 1	246.5° E
	GTE G–Star	256.9° E
	GE Satcom K2	279.1° E
	GTE Spacenet 2	291.0° E
	SBTS 1	295.0° E
	INTELSAT VA F11	332.5° E
	INTELSAT VA F10	335.5° E
	Telecom 1–B	354.9° E

1986	Raduga 19	45.8° E
	Gorizont 13	89.2° E
	PRC 18 China	104.7° E
	BS–2B Japan	110.0° E
	GTE G–Star	255.0° E
	SBTS 2	289.9° E
	Raduga 18	335.0° E
	Gorizont 12	346.2° E
1987	ECS 4	9.9° E
	Raduga 20	85.3° E
	Ekran 16	99.1° E
	Palapa 2–B	113.0° E
	CS–3A	135.0° E
	Gorizont 14	139.3° E
	ETS 5	150.0° E
	Volna 1–A	335.0° E

Appendix 2
US Satellite Footprints

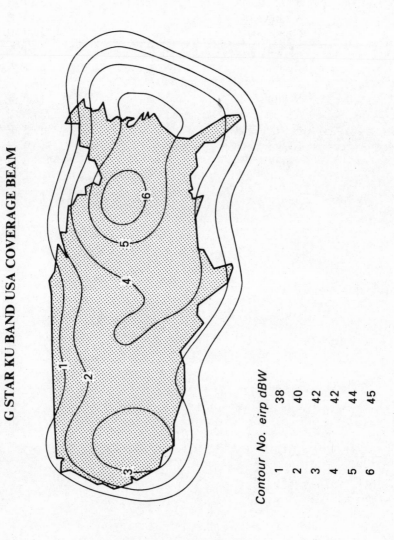

G STAR KU BAND USA COVERAGE BEAM

Contour No.	eirp dBW
1	38
2	40
3	42
4	42
5	44
6	45

SPACENET 1 KU BAND USA COVERAGE

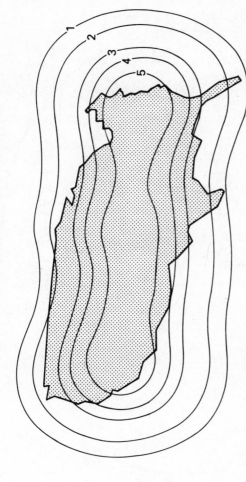

Contour No. eirp dBW

1	35.8
2	38.8
3	41.8
4	42.8
5	43.8

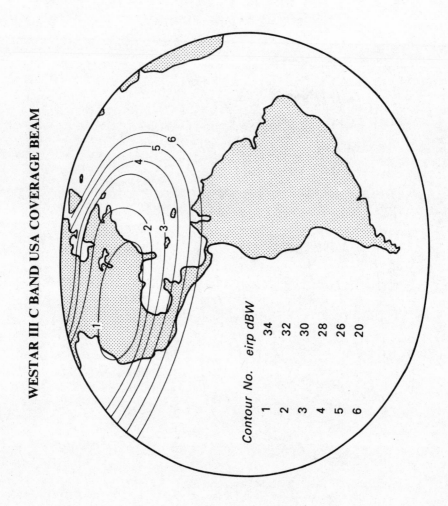

WESTAR III C BAND USA COVERAGE BEAM

Contour No.	eirp dBW
1	34
2	32
3	30
4	28
5	26
6	20

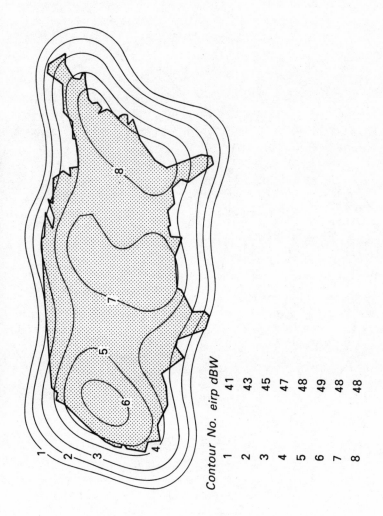

SATCOM K1 KU BAND USA COVERAGE BEAM

Contour No.	eirp dBW
1	41
2	43
3	45
4	47
5	48
6	49
7	48
8	48

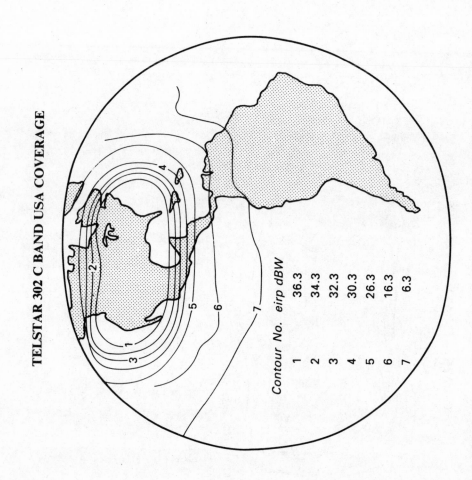

TELSTAR 302 C BAND USA COVERAGE

Contour No.	eirp dBW
1	36.3
2	34.3
3	32.3
4	30.3
5	26.3
6	16.3
7	6.3

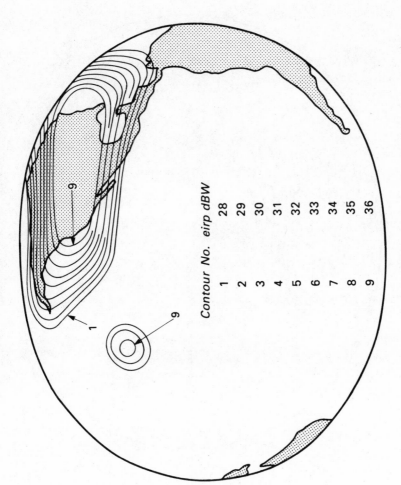

GALAXY 1 C BAND TOTAL USA COVERAGE

Contour No.	eirp dBW
1	28
2	29
3	30
4	31
5	32
6	33
7	34
8	35
9	36

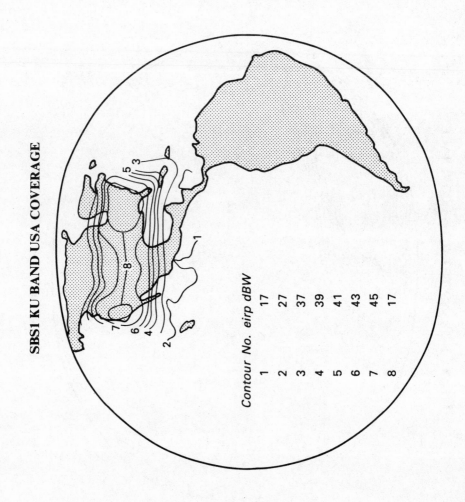

SBS1 KU BAND USA COVERAGE

Contour No.	eirp dBW
1	17
2	27
3	37
4	39
5	41
6	43
7	45
8	17

Appendix 3
Bibliography

CHAPTER 1

Arabsat C Band Earth Station Peformance Characteristics Document/T/ES-1.

'A review of the development of the INTELSAT system'. G.C. Hall and P.R. Moss. *POEEJ* vol. 71, October 1978.

EUTELSAT Document. *The European telecommunications satellites.*

EUTELSAT II Earth Station Seminar, January 1987.

'Insat ground segment'. U.V. Nayak. *IETE Tech Review* vol. 2 no. 6 1985.

INTELSAT. The global telecommunications co-operative.

'Satellite broadcasting in Australia'. M. Bridle. *IEEE Transaction on Broadcasting* vol. 34 no. 4, December 1988.

Satellite communications. J. Wilson. Telecom Australia.

Satellite communications networks. M. Nouri. MCSL.

'SATNET packet data transmission'. Palmer, Kaiser, Rothschild and Mills. *COMSAT Technical Review* vol. 12 no. 1, Spring 1982.

'The evolution of the INTELSAT V system and satellite'. R.J. Eaton and R.K. Kirby. *POEEJ* vol. 70, March/April 1977.

World satellite almanac 1985. Mark Long. CommTek Publishing Company.

CHAPTER 2

'A survey of, ionospheric effects upon Earth-space radio propagation'. Lawrence, Little and Chivers. *IEEE Proceedings*, January 1963.

'Atmospheric attenuation in satellite communications'. W. Holzar. *Microwave Journal*, March 1965.

'Cross polarisation in satellite and Earth station antennas'. Samier I. Ghobrial. *IEEE Proceedings* vol. 65 no. 3, March 1977.

'Earth footprints of satellite antennas'. Jacobs and J.M. Stacey. *IEEE Transactions AES*, March 1971.

'Geostationary satellite orbital geometry'. L.P. Yeh. *IEEE Transactions on Communications* vol. Com-20 no. 2, April 1972.

'Prediction of attenuation on satellite-earth links in the European region'.

Professor P. A. Watson *et al. IEE Proceedings* vol. 134 PEF, October 1987.

'Propagation phenomena affecting satellite communications'. R. K. Crane. *IEEE Proceedings* vol. 59 no. 2, February 1971.

Signal attenuation due to neutral oxygen and water vapour, rain and clouds. Andre Benoit.

'System implications of 14/11GHz path depolarization'. D. V. Rogers and J. E. Allnutt. *International Journal of Satellite Communications* vol. 4 1–11, 1986.

'Ultimate communication capacity of the geostationary satellite orbit'. J. K. S. Jowett and A. K. Jefferies *IEE Proceedings* vol. 116 no. 8, August 1969.

CHAPTER 3

'B-MAC: a transmission standard for pay DBS'. K. Lucas. *SMPTE Journal,* November 1985.

Digital communications by satellite. J. J. Spilker. Prentice Hall, 1977.

'Digital speech interpolation'. S. J. Campanella. *COMSAT Technical Review* 6 (1), 1976.

D2 – MAC/packet system specification. Ministry of Industry, SERICS, 32 Rue Guersant 75017, Paris, September 1985.

'D2 – MAC/packet system transmission and vision encoder sub-system'. G. Duvic and H. Lebloch. *Communication and Transmission* no. 2, 1988.

'Modulation and coding'. P. G. Farrell and A. P. Clark. *International Journal of Satellite Communications* vol. 2 287–304, 1984.

Multichannel communications systems and white noise testing. M. J. Tant. Marconi Instruments.

PCM and digital transmission systems. F.F.E. Owen. McGraw Hill.

'Specification of the systems of the MAC/packet family'. EBU Document Tech. 3258 (1986).

Television engineering broadcast, cable and satellite Part 1 and 2 RTS.

'The ALOHA system'. N. Abrahson. *AFIPS Conference Proceedings* vol. 37, 1970.

'The Viterbi algorithm'. G. D. Forney. *IEEE Proceedings* vol. 61 no. 3, March 1973.

'Viterbi decoding for satellite and space communications'. Heller and Jacobs. *Transactions on Communications Technology* vol. COM-19 no. 5, October 1971.

CHAPTER 4

CCITT yellow book. Part III Rec G791–G794.

'Earth-terminal design benefits from MMIC technology'. N. K. Osbrink, MSN and CT, August 1986.

'Echo control'. Walter J. Morgan. *Satellite Communications*, July 1985.

'Elliptical offset Gregorian antenna for a transportable Earth station'. R. I. Henderson. *GEC Journal of Research* vol. 2 no. 3, 1984.

'Intermodulation effects in limiter amplifier repeaters'. Bond and Meyer. *IEEE Transactions on Communications Technology* vol. COM-18, April 1970.

'Lobes, sidelobes and cross polarisation caused by feed-support struts'. Kildal, Olsen and Anders. *IEEE Transactions on Antenna and Propagation* vol. 36 no. 2, February 1988.

'Low-cost flat-plate array with squinted beam for dbs reception'. *IEE Proceedings* vol. 134 Pt H no. 6, December 1987.

'Tracking systems for satellite communications'. Hawkins, Edwards and McGecham. *IEE Proceedings* vol. 135 Pt F no. 5, October 1988.

'Wideband high power TWTs designed for telecommunications'. Kuntzmann, Smith and Tikes. *MSN and CT*, March 1986.

CHAPTER 5

INTELSAT Documents:

IESS–301 Rev 1. Performance characteristics for FDM/FM telephony carriers.

IESS–302 Rev 3. Performance characteristics for CFDM/FM telephony carriers.

IESS–305 SCPC/CFM. Performance characteristics for the vista service.

IESS–401. Performance requirements for intermodulation transmitted from INTELSAT Earth stations.

IESS–402 Rev 1. E/S adjustment factors.

CHAPTER 6

'A uni-directional satellite paging system'. I. E. Casewell, I. C. Ferebee and Professor M. Tomlinson. *Journal IERE*, May 1988.

Global positioning system vol. 1, 2 and 3. Institute of Navigation.

INMARSAT aeronautical communication services system summary. INMARSAT, May 1987.

'ITU activities in the field of mobile services'. *Telecommunication Journal* vol. 54 VI/1987.

MSAT X: a technical introduction and status report. JPC Publication, 88–12, April 1988.

Navstar GPS system characteristics. STANAG 4294 Draft Issue H.

'Search and rescue by satellite'. C. Bulloch. *Interavia* 3, 1987.

Specification of COSPAS/SARSAT 406 MHz Distress Beacons C/S T.001. Iss, 1 April 1986.

'Understanding signals from Glonass navigation satellites'. Dale, Daley and Kitching. *International Journal of Satellite Communications* vol. 7 no. 11–22, 1989.

'The INMARSAT System'. C. B. Wooster. *Communications and Broadcasting* vol. 7 no. 2.

CHAPTER 7

'An Earth station for the EUTELSAT Multi-Service System'. F. L. van den Berg. *Phillips Communication Review* vol. 43 no. 1, 1985.

Apollo system requirements document ESA SP1068. EVR 9413 GN, August 1984.

'Comparison of costs of TDMA and IDR' J. Colby and A. K. Kwan. *International Journal of Satellite Communications* vol. 6, 1988.

'Experimental IBS Small Dish Earth Station'. Y. Hirata *et al. IEEE Conference on Communications* vol. 1, 1987.

'FASTCOM: small rural satellite earth stations'. J. Salomon. *Communications and Transmission*, 1986.

INTELSAT Earth Station specification IESS 308 IDR Performance.

INTELSAT Earth Station specification IESS 601 Standard G.

INTELSAT Earth Station specification IESS 602 Standard Z.

INTELSAT Earth Station specification IESS 309 IBS Service.

'Ku Band Satellite Digital Transmission System'. W. E. McGane. *International Journal of Satellite Communications* vol. 3, 1985.

'Local satellite communications system for high reliability'. H. Kawaharce. *JTR*, January 1987.

CHAPTER 8

IEE colloquium on small terminal satellite communications 1986. IEE.

'Ku Band satellite data networks using VSATs'. D. Raychaudhari and K. Joseph. *International Journal of Satellite Communications* vol. 5, 1985.

Small Earth station symposium and exposition, May 1987.

CHAPTER 9

'A conditional access system for direct broadcasting by satellite'. S. M. Edwardson. *JIERE* vol. 55 no. 11/12, 1985.

'DBS in the USA'. *IEEE Communications Magazine* vol. 22 no. 3, March 1984.

DBS, SMATV and cable in technical harmony. Proceedings of Cable 85 Conference, 1985. R. J. Seacombe.

'Direct broadcasting by satellite'. C. B. Wooster. *Communications and Broadcasting* vol. 8 no. 1.

'Satellite broadcasting in Australia'. M. Bridle. *IEEE Transactions on Broadcasting* vol. 34 no. 4, December 1988.

'Satellite FM television channel bandwidth'. C. A. Siocos. *Space Communications and Broadcasting 2* (1984).

'TV distribution by satellite'. B. Ackroyd. *Communications and Broadcasting* vol. 9 no. 3, 1985.

CHAPTER 10

CCIR Space Research and Radio Astronomy Report 693–2 and *Report 535–3.*
Earth Observing System IEEE Transactions on Geoscience and Remote Sensing vol. 27 no. 2, March 1989.

'Looking down for money'. Jaques and Lopez. *Space Markets*, Summer 1986.

'Meteosat ranging concept'. Otter and Drewes. *ESA Journal* vol. 12, 1988.

'SAR images of moving targets'. Freeman and Currie. *GEC Journal of Research* vol. 5 no. 2, 1987.

'Sensors behind the spot'. D. King. *Satellite Communications*, August 1988.

'The ERS-1 synthetic aperture radar and scatterometer'. Brooks, Joyce, Sawyer and Smith. *GEC Journal of Research* vol. 3 no. 2, 1985.

'The Meteosat Programme'. R. Tessler. *ESA Journal.*

CHAPTER 11

'Competition between transatlantic fibre optic cables and satellites'. *Telecommunications Journal* vol. 54 X11, 1987.

'Cost comparison of microwave, satellite and fibre optic systems'. P. Polishuk, R. Geunther and J. Lawlor. *Telecommunications Journal* vol. 54 no. 11, 1987.

'Factors affecting Fibre optic Transmission quality'. T. Kimura. *Journal of Light Wave Technology* vol. 6 no. 5, May 1988.

'Optical FDM transmission Technique Nosu. Toba and Iwashita'. *Journal of Light Wave Technology* vol. LT–5 no. 9, September 1987.

Proceedings of international conference on satellite and optical fibre. Fibresat 86, Vancouver 1986.

'Satellite Communications'. *Aviation Week and Space Technology*, December 1986.

'Satellites versus fibre optic cables'. J. Cummins, J. M. Cummins, Lemus and Reyna. *Journal of Satellite Communications*, March 1985.

'Wavelength division multiplexing over optical fibre'. Hanson and Peers. *Ericsson Review* no. 2, 1988.

World Satellite Communications

CHAPTER 12

'20–30GHZ Band Earth Station for Domestic Satellite Communications'. *Electronics and Communication in Japan* vol. 63 no. 8, 1980.

20/30GHz *E/S Components for SDCS*. T. Inove, Y. Yamada and F. Kawashima.

'Contribution of satellite information to operational weather forecasting'. J. Reiff. *International Journal Remote Sensing* vol. 9 nos. 10 and 11, 1988.

'Dynamic Power in Space'. W. Miskell. *Space* vol. 1 no. 3, December–February 1986.

IEEE Journal on Selected Areas in Communications vol. SAC–5 no. 4, May 1987.

'Ion thrusters advance'. A. Wilson. *Space* vol. 3 no. 3, July–August 1987.

'Satellite communications'. S. J. Campanella. *Space* vol. 3 no. 4, September–October 1987.

'Satellite Digital Communications Service (SDCS)'. Y. Morihiro, H. Nakashima and S. Kato. *Review of Electrical Communication Laboratories* vol. 35 no. 2, 1987.

Satellite overview. Tetsuo Goh Advance Technical Report. Mitsubishi.

TDMA System for SCDS: Review of the Electrical Communications Laboratories vol. 35 no. 2, 1987. S. Kato, M. Morikura and M. Umehira.

'The large telecommunications satellite Olympus'. J. H. Paul. *ESA Journal*.

Appendix 4

Glossary of Satellite Communication Acronyms

ABC	American Broadcasting Company
ACI	Adjacent channel interference
ACME	Analogue circuit multiplication equipment
ADC	Analogue to digital converters
ADM	Adaptive delta modulation
ADPCM	Adaptive differential pulse code modulation
AFC	Automatic frequency control
AFRTS	American Forces Radio & Television Services (USA)
AGC	Automatic gain control
AM	Amplitude modulation
AMI	Active microwave instrument
ANBFM	Adaptive narrow band frequency modulation
AOR	Atlantic Ocean region
APC	Adaptive predictive coding
ARINC	Aeronautical Radio Incorporated
ARQ	Automatic repeat request
ASAR	Advanced synthetic aperture radar
ASEAN	Association of South East Asian Nations
ATC	Adaptive transform coding
ATS	Applications technology satellite
ATSR	Along track scanning radiometer
AUSSAT	Australian National Satellite System Operating Company
AVDM	Analogue variable delta modulation
AVHRR	Advanced very high resolution radiometer
BCD	Binary coded decimal
BCH	Bose–Chaudhuri–Hocquengherm
BER	Bit error rate
BPF	Band pass filter
BPSK	Binary phase shift keying
BRAZILSAT	Brazilian satellite system
CATA	Community Antenna Television Association

CATV	Community antenna television
CBC	Canadian Broadcasting Corporation
CBS	Columbia Broadcasting System
CCD	Charge coupled device
CCI	Co-channel interference
CCIR	International Radio Consultative Committee
CCITT	International Telegraphy & Telephony Consultative Committee
CDM	Code delta modulation
CDMA	Code division multiple access
CEPT	Conférence Européenne des Administration des Postes et des Télécommunications
CES	Coast earth station
CFM	Companded frequency modulation
CNES	Centre National d'Etudes Spatiale
CNN	Cable News Network (US)
COMSAT	Communication Satellite Corporation
CONUS	Continental United States
CPSK	Coherent phase shift keying
DAMA	Demand assigned multiple access
DATTS	Data acquisition telecommand and tracking station
dbs	Direct broadcasting by satellite
DCME	Digital circuit multiplication equipment
DCP	Data collection platform
DEDM	Dolby enhanced delta modulation
DEPSK	Differentially encoded phase shift keying
DPCM	Differential pulse code modulation
DSI	Digital speech interpolation
EBU	European Broadcasting Union
ECS	European communication satellite
eirp	Equivalent isotropic radiated power
ELT	Emergency locator transmitter
EPIRB	Emergency position-indicating radio beacon
ESA	European Space Agency
ESC	Engineering service circuits
FCC	Federal Communication Commission
FDM	Frequency division multiplex
FDM/FM	FDM/frequency modulation
FEC	Forward error correction
FET	Field effect transistor
FFSK	Fast frequency shift keying

FM	Frequency modulation
FMFB	Frequency modulation feedback
FSK	Frequency shift keying
FSS	Fixed satellite service
GEOS	Geodetic earth orbiting satellite
GOES	Geostationary operational environmental satellite
GPS	Global positioning system
GRIN	GRaded INdex miniature lens
GSO	Geostationary orbit
HACBSS	Homestead & Community Broadcasting Satellite Services
HPA	High power amplifier
HRC	Hypothetical reference circuit
IBM	International Business Machines
IBS	Intelsat Business Services
ICAO	International Civil Aviation Organization
IDR	Intermediate data rate
IFRB	International Frequency Registration Board
INMARSAT	International Maritime Satellite Organization
INSAT	Indian National Satellite System
INTELSAT	International Telecommunications Satellite Organization
IOR	Indian Ocean region
IPA	Intermediate power amplifier
ISDN	Integrated services digital network
ITU	International Telecommunications Union
LAN	Local area network
LASER	Light amplification by simulated emission of radiation
LEO	Low earth orbit
LHCP	Lefthand circular polarisation
LNA	Low noise amplifier
LNB	Low noise block
LNC	Low noise converter
LPC	Linear predictive coding
LPF	Low pass filter
LSI	Large scale integration
LUT	Local user terminal
MA	Multiple access
MAC	Multiplexed analogue components

MASER	Microwave amplification by simulated emission of radiation
MATV	Master antenna television
MCC	Mission control centre
MCS	Maritime Communication System (IS IV)
METEOSAT	Meteorological satellite
MMIC	Monolithic microwave integrated circuit
MSK	Minimum shift keying
MSS	Mobile satellite service
MTBF	Mean time between failures
MTTR	Mean time to repair
NASA	United States National Aeronautics & Space Administration
NAVSTAR/GPS	Navstar Global Positioning System
NBC	National Broadcasting Company
NBFM	Narrowband frequency modulation
NBPCM	Non-binary PCM
NCS	Network co-ordination station
NOAA	National Oceanic Atmospheric Administration
NOC	Network Operations Centre
NTSC	National Television Standards Commission (US)
OMT	Orthogonal-mode transducer
OTS	Orbital test satellite
PAL	Phase alternation by line TV encoding system
PAM	Pulse amplitude modulation
PCM	Pulse code modulation
PD	Phase detector
PFD	Power flux density
PLL	Phase lock loop
PM	Phase modulation
POPSAT	Precise orbit positioning satellite
POR	Pacific Ocean region
PSK	Phase shift keying
PSTN	Public switched telephone network
PTT	Post Telephone & Telegraph
QPSK	Quadrature phase shift keying
RADARSAT	Radar satellite
RARC	Regional Administrative Radio Conference
RF	Radio frequency

RFI	Radio frequency interference
RHCP	Right hand circularly polarised
RPFD	Received power flux density
RS	Reed Salomon
SAR	Search and rescue
SARSAT	Search and rescue satellite
SBS	Satellite business systems
SCPC	Single channel per carrier
SECAM	Sequence Coleur à Mémoire (TV encoding system from France)
SES	Ship earth station
SMATV	Satellite master television
SNR	Signal to noise ratio
SPADE	Single channel per carrier DAMA
SPOT	Système Probatoire d'Observation de la Terre
SSPA	Solid state power amplifier
SSTDMA	Satellite switched time division multiple access
TCM/DPSK	Trellis coded modulation/differential phase shift keying
TDM	Time division multiplex
TDMA	Time division multiple access
TDRS	Tracking and data relay satellite
TT&C	Tracking, telemetry & command
TVRO	Television receive only
TWT	Travelling wave tube
UHF	Ultra-high frequencies
UN	United Nations
VCO	Voltage controlled oscillator
VHF	Very high frequency
VLSI	Very large scale integration
VOA	Voice of America
VSAT	Very small aperture terminals
WARC	World Administrative Radio Conference
WDM	Wavelength division multiplexing
WMO	World Meteorological Organization

Glossary of Satellite Communication Terms

Adjacent channel interference Interference between two transmission channels adjacent to one another.

Amplifier Input/output device which increases the level of a signal which may be voltage, current, power or light.

Amplitude modulation A type of modulation in which the carrier wave amplitude varies with the frequency of the modulating signal.

Analogue transmission A type of transmission where the information is of a continuously variable form.

Antenna A device used to gather or send signals to and from a satellite.

Aperture The area of an antenna that can 'see' the signal from the satellite.

Apogee The maximum altitude on a satellite orbit.

Asynchronous operation A method of data transmission in which data is sent in a stream.

Attenuator An active or passive component that reduces the level of a signal between two points in the link.

Axial ratio A parameter associated with polarisation circularity and cross-polarisation performance of an antenna.

Azimuth angle The angle between the antenna beam and meridian plane, measured along the horizontal plane.

Bandwidth The effective frequency band over which a device, such as an amplifier operates. The points at which bandwidth is measured are usually at the frequencies where the signal suffers a loss of 3 dB relative to the minimum loss point.

Baseband The frequency band which contains the basic information prior to modulation and after demodulation.

Binary coded information Information which is represented by binary numbers to allow transmission to take place.

Bit error rate (BER) A data transmission is composed of discrete bits of information which can be corrupted and received incorrectly, giving errors. The ratio of the number of bits incorrectly sent to the total number sent is the bit error rate.

Carrier to noise (C/N) Ratio of satellite carrier level to noise level in a given channel.

Cassegrain Dual reflector antenna with paraboloid main reflector and a hyperboloid sub-reflector.

Circular polarisation Transmission mode in a circular rotating pattern.

Code division multiple access A multiple access system that uses spread spectrum method of modulation.

Codec An equipment or device which contains both encoder and decoder.

Coherent detection A detection method used in suppressed carrier/sideband systems that re-constitutes the reduced carrier.

Common carrier A public or private body able to provide a general telecommunication service to private individuals and businesses.

Companding Signal compression to reduce noise level in signal.

Conical scan A form of tracking which uses a rotating feed.

Cross talk A condition where signals in one channel enter into a channel adjacent to it. Cross talk can be intelligible or non-intelligible.

Cyclic code A coding system used for error detection.

Data channel A transmission channel used to carry data between two locations.

Decibel (dB) A unit of measure for the level of a signal.

Delay distortion Signal impairment in a circuit due to the variation in the propagation time at different frequencies.

Delay time For satellites this refers to the delay experienced by the earth station signals on their path to the satellite.

Demultiplexer An equipment or device used to separate out individual signals that were originally combined in the same transmission channel.

Direct broadcast Refers to satellites that broadcast image and data direct to home or office.

Distortion A signal impairment caused by the transmission link, which can be linear or non-linear.

Down conversion An equipment used to translate a frequency down to a lower one.

Down link The link between satellite and earth station.

Earth station A system of transmit/receive equipment, used with a parabolic antenna, for sending signals to and receiving signals from the satellite.

Echo distortion Signal impairment due to part of the transmission being delayed and later recombined with the main signal.

Elevation The angle between the main antenna and the horizontal plane.

Encryption The manipulation of a composite signal to render it unintelligible without the use of a receive decoder.

Energy dispersal The dispersal of carrier power across the transmission bandwidth by the use of a triangular waveform to reduce inter channel interference.

Figure of merit The measure of quality of a satellite earth station.

Focal length The distance from centre of antenna reflector to the feed.

Footprint The satellite beam area of coverage on the earth's surface.

Frequency re-use The transmission of signals in the same frequency band with discrimination being obtained by the use of different polarisations.

Geostationary satellite A satellite which is positioned in orbit above the earth at a height such that its speed is synchronised to the earth's rotation and thus appears stationary from the earth's surface.

Gregorian antenna An antenna which uses a parabolic main reflector and a concave ellipsoidal sub-reflector.

Group delay distortion An impairment caused by different frequencies of transmission being subjected to different delay causing distortion of signals.

Guard band A portion of the frequency band kept unused to separate different channels and so prevent unwanted signals entering the channels.

Half transponder operation A method of transmitting two TV signals through one 36 MHz transponder, though at lower power and deviation.

Head end The central distribution point for cable television systems.

Hub station The central control earth station for a group of VSATs.

Intermodulation The phenomenon found when non-linearity in a transmission system produces a modulating effect upon an adjacent carrier.

Isotropic antenna A reference omni-directional antenna against which other antenna performance is measured.

Jitter The process in which the transitions of a digital signal advance or retard from their correct positions. This phenomenon is a major cause of errors in a transmission system.

Kelvin The unit of absolute temperature used in noise measurement. The point of absolute zero is $-273°$ Celsius.

Kilobit A unit of one thousand binary digits.

Link budget The assessment of all the factors affecting the signal transmitted to and from the satellite, with the view of determining the system performance in terms of carrier/noise ratio.

Low noise amplifier An amplifier placed at the front of a satellite receiving system to give the best system signal to noise available.

Meridian A line passing through one's own position on the earth and the earth's axis.

Modem A modulator/demodulator unit.

Modulation The processing of a carrier wave with the baseband information that needs to be transmitted to ensure intelligible transmission over longer distances. The processing can involve variations in amplitude, phase, frequency or combinations of these.

Monopulse A type of tracking system using multiple feed horns.

Multiple access The system whereby more than one user can access the transponder at the same time. This access can be in the frequency, time or space domain.

Multiplexer An equipment or device that combines a number of message signals on to a single tributary.

Noise Any signal that interferes with the transmitted signal.

Noise figure A measure of the noise in a system, relative to a zero noise system, measured in dB.

Noise temperature A measure of the noise of a system defined as the absolute temperature of a resistive source delivering equal noise power, measured in degrees Kelvin.

Offset antenna A front-fed antenna whose feed is not directly on the axis of the reflector but is offset to reduce signal blockage.

Orthogonal At right angles.

Paraboloid A parabola of revolution.

Phase The relative position of two waveforms on a time axis.

Polarisation An electromagnetic wave property involving signal rotation.

Protocol The set of rules which define the way in which information can flow in a system.

Quantisation A process used in pulse amplitude modulation involving the sampling of an analogue signal to transform it into digital form.

Random noise A form of interference that is known as white or gaussian noise.

Redundancy An arrangement of earth station equipment to ensure that if failure occurs working equipment can be inserted to take over the failed function.

Satellite A communication centre placed in orbit around the earth allowing communication to be made over longer distances than by most conventional means and the switching of the signal to different destinations.

Shaped antenna reflectors A method of modifying antenna characteristics by making the reflector shape different to its classical primary form.

Sideband A signal generated by a modulation process above and below the carrier signal at frequencies determined by the modulating signal.

Sidelobes The off-axis response of an antenna.

Solar outage The disruption of a satellite operation because of radiated noise from the sun. This occurs either when the satellite antennas are looking at the sun or the sun passes within the field of view of the satellite antenna.

Spread spectrum A signal coding technique that provides secure communication by spreading the signal energy across a wide bandwidth.

Synchronisation A process of timing all network elements from the same source to allow simple operation.

Teleconferencing A communication system that links together people at different locations by use of satellite links. It can involve both audio and video methods.

Threshold The point at which the relationship between carrier/noise and signal/noise ratio in an FM system deviates from the linear by 1 dB.

Threshold extension The extending of the value of C/N ratio at which threshold occurs.

Transponder The equipment in a satellite that receives a signal from the ground station, translates it to the downlink frequency and transmits the signal.

Truncation An effect caused by excessive filtering of a signal that reduces the intelligibility of a signal.

Uplink The transmission of signals from an earth station up to the satellite.

Visible arc The arc of the geostationary orbit over which the satellite is visible to the earth station.

Voice circuit A circuit used to carry speech, normally having a bandwidth of 300–3400 Hz,

White noise A form of interference on transmission systems caused by basic movement of electrons.

Index